T0184107

A GUIDE TO PROBLEMS IN MODERN ELECTROCHEMISTRY

1: Ionics

A GUIDE
TO PROBLEMS
IN MODERN
ELECTROCHEMISTRY

1: Ionics

Maria E. Gamboa-Adelco
Superior, Colorado

and

Robert J. Gale
Louisiana State University
Baton Rouge, Louisiana

Kluwer Academic/Plenum Publishers
New York, Boston, Dordrecht, London, Moscow

ISBN 0-306-46668-6

©2001 Kluwer Academic/Plenum Publishers, New York
233 Spring Street, New York, New York 10013

http://www.wkap.nl/

10 9 8 7 6 5 4 3 2 1

A C.I.P. record for this book is available from the Library of Congress

All rights reserved

No part of this book may be reproduced, stored in a retrieval system, or transmitted in
any form or by any means, electronic, mechanical, photocopying, microfilming, recording,
or otherwise, without written permission from the Publisher

Printed in the United States of America

To all students of Electrochemistry

PREFACE

The main goal when writing the textbook *Modern Electrochemistry* –in its original version of 1970 (by Bockris and Reddy), and later in its second edition of 1998 (by Bockris, Reddy and Gamboa-Aldeco)– was to present, in a lucid way, the complex and multidisciplinary area of Electrochemistry. The aim was to start at a low level and increase the complexity of concepts, theories, and their applications. The result is a detailed presentation of topics such as ion-solvent and ion-ion interactions (Volume 1: Ionics), and thermodynamics and kinetics of electrodic processes (Volumes 2A and 2B: Electrodics). In this effort, the involvement of various other disciplines –such as mathematics, biology and engineering– were considered as well as numerous applications, such as fuel cells, production of metals, batteries, soil remediation and photo-splitting of water, to mention a few.

The textbook is addressed to anyone interested in learning Electrochemistry, with occasional previews of advancing frontier topics. It is written for students or anyone who, because of their work or personal interests, would be presented with the challenge of learning Electrochemistry. Since the goal was to *teach*, the authors followed the format of an excellent textbook, *Physical Chemistry*, written by Peter Atkins (W.H. Freeman). Therefore, in the second edition of Modern Electrochemistry, exercise headings addressed to assimilate the acquired knowledge were included at the end of each chapter. Thus, any student willing to improve his/her understanding of this field could be challenged by solving the many exercises found through the book.

It has been always an incentive for students to find whether his/her efforts to solve exercises give correct results, or to find tips for problems that he/she finds more difficult. These are the main reasons for the appearance of the present book. As part of the textbook *Modern Electrochemistry 1: Ionics*, *A Guide to Problems in Modern Electrochemistry: Part 1 Ionics* compiles many of the solutions to the exercises and problems presented in the text, as well as many new problems. During its preparation, the authors wanted to emphasize three points; first, the variation in *complexity* of the exercises, which allows the student to acquire the knowledge in little steps; second, the *numerous* exercises and problems that give the student plenty of practice to assimilate the concepts in the book; and third, the *variety* of topics presented in the book, which covers fundamental concepts through applications in modern technology.

The exercises in each chapter have been divided into three categories. The first category contains simple *Exercises*, which can be completed with a basic understanding of the related text. These *Exercises* allow for practice in using the chapter's equations and should take about 15 minutes each to solve. *Exercises* represents 37% of the total number of exercises and problems, with an average of 30 exercises per chapter. In the second category, the degree of complexity increases. This is not because of the length of the answers, but because of a more profound understanding of the concepts is required. Because of their higher complexity, the *Problems* require more time to solve, but can each be completed within one hour. There are about 44 of these *Problems* in each chapter, representing 60% of the total number. Finally, the third category includes tasks that are more difficult. To solve these *Micro-Research Problems*, a further effort of the student is needed. This may imply research on his/her own for data or deeper concepts in reference or more advanced texts. There is only one of these problems included in each chapter.

At this point it is important to stress that, in addition to the efforts of the authors, more than thirty Electrochemistry professionals contributed to the preparation of the problems. Thanks to them, the authors, with substantial assistance from J. O'M. Bockris, were able to compile and organize more than 500 exercises and problems for the whole book, including those in **A Guide to Problems in Modern Electrochemistry: Part 2 Electrodics**. Just writing this many questions –without considering their solutions, complexity, and variety of topics– constitutes a tremendous effort. The authors would like to thank each one of the participants of this volume who make this book an asset for the student of Electrochemistry: Bock, Bockris, Casidar, Constantinescu, Contractor, Herbert, Kim, Mussini, Tejada, Trassati, and Xu.

Finally, one of the authors (Gamboa-Aldeco) thanks Prof. John O'M. Bockris, for his encouragement, guidance, insistence, and recompilation of the problems of this book.

CONTENTS

CHAPTER 3

IONIC LIQUIDS

Exercises
Problems
Micro-Research Problems

INDEX

CONTENTS

CHAPTER 5

CHAPTER 1

NOMENCLATURE

For each symbol the appropriate SI unit is given. Multiples and submultiples of these units are equally acceptable, and are often more convenient. For example, although the SI unit of concentration is **mol m^{-3}**, concentrations are frequently expressed in **mol dm^{-3} (or M)**. The abbreviation *SI units* comes from *Système International d'Unités*, a system developed by the General Conference on Weights and Measures based on the *mksa* (meter-kilogram-second-ampere) system.

Symbol	Name	SI unit	Other units frequently used

GENERAL

Symbol	Name	SI unit	Other units
A	area of an electrode-solution interface	m^2	
a_i	activity of species i		
c_i	concentration of species i	mol m^{-3}	M, N
c^o	bulk concentration	mol m^{-3}	M, N
E	energy	J	
F	force	N	
$g_{i,j}$	radial pair distribution function		
v	frequency	Hz, s^{-1}	
m	mass	kg	
n	number of moles		
N	number of molecules		

Symbol	Name	SI unit	Other units frequently used
P	pressure	Pa	atm
r, d, l	distance	m	Å
T	thermodynamic temperature	K	°C
t	time	s	
U	potential energy	J	
V	volume	m^3	
v	velocity	$m\,s^{-1}$	
W	work	J	
x_i	molar fraction of species i		
$\bar{\mu}$	reduced mass		
ρ	density	$Kg\,m^{-3}$	
λ	wavelength	m	
$\tilde{v}$	wavenumber	m^{-1}	
θ	angle	°	
γ	surface tension	$N\,m^{-1}$	$dyn\,cm^{-1}$

ION AND MOLECULE-RELATED QUANTITIES

a	distance of closest approach	m	Å
q	Bjerrum parameter	m	
z_i	charge number of an ion i		
α_i	polarizability of species i	$m^3\,molecule^{-1}$	
β	compressibility	Pa^{-1}	
μ	dipole moment	C m	D (debye) esu cm
κ^{-1}	Debye-Hückel reciprocal length	m^{-1}	

THERMODYNAMICS OF A SIMPLE PHASE

$E_a, E^{\neq}$	energy of activation	$J\,mol^{-1}$	
ΔG	relative molar Gibbs free energy	$J\,mol^{-1}$	
$\Delta H, L$	relative molar enthalpy	$J\,mol^{-1}$	
K	equilibrium constant of the reaction		
Q	charge	C	
ΔS	relative entropy	$J\,K^{-1}\,mol^{-1}$	eu

Symbol	Name	SI unit	Other units frequently used
$\vec{X}$	electric field	V m^{-1}	
μ_i	chemical potential of species i	J mol^{-1}	
$\overline{\mu}_i$	electrochemical potential of species i	J mol^{-1}	
ρ	charge density	C	
ε	dielectric constant		
ψ	electrostatic potential between two points	V	

ACTIVITIES IN ELECTROLYTIC SOLUTIONS AND RELATED QUANTITIES

$a_\pm$	mean activity		
$\gamma_b\, f_i$	activity coefficient of species i		
$f_\pm\, \gamma_\pm$	stoichiometric mean molar activity coefficient		
I	ionic strength		

MASS TRANSPORT

D_i	diffusion coefficient of species i	m^2 s^{-1}	cm^2 s^{-1}
$\vec{J}_i$	flux density of species i	mol m^{-2} s^{-1}	
v_d	drift velocity	m s^{-1}	
η	viscosity	g cm^{-1} s^{-1}	poise

CHARGE TRANSPORT PROPERTIES OF ELECTROLYTES

q_i	charge of species i	C	
R	resistance of the solution	Ω	
$1/R$	conductance	S or Ω^{-1}	
σ	specific conductivity	S m^{-1}	
Λ_m	molar conductivity of an electrolyte	S m^2 mol^{-1}	

Symbol	Name	SI unit	Other units frequently used
λ_i	ionic conductivity or ionic species i	S m^2 mol^{-1}	
Λ	equivalent conductivity	S m^2 eq^{-1}	
t_i	transport number of ionic species i in an electrolytic solution		
$\bar{u}_{abs}$	absolute mobility	m s^{-1} N^{-1}	
u_i, $(u_{conv})_i$	conventional (electrochemical) mobility of species i	m^2 V^{-1} s^{-1}	

KINETIC PARAMETERS

Symbol	Name	SI unit
I	electric current	A
j	current density	A m^{-2}
$\vec{k}, \overleftarrow{k}$	rate constants	

STATISTICS AND OTHER MATHEMATICAL SYMBOLS

Symbol	Name
P_r	probability
$<x>$	average value of variable x
$<x^2>$	mean square value of variable x
x_{rms}	root-mean-square value of variable x
θ	fraction number
$erf(y)$	error function
$x_{\pm}$	mean value of variable x
$\Gamma(n)$	Gamma function

Useful Unit Conversion Factors

Potential	Length	Volume	Mass	Force	Energy
1 V	1 m	1 m^3	1 g	1 N	1 J
1 J C^{-1}	100 cm	1000 dm^3	10^{-3} kg	1 kg m s^{-2}	1 kg m^2 s^{-2}
1 A Ω	1000 mm	1000 liters	1000 mg	10^5 dynes	1 N m
1 C Ω s^{-1}	10^6 μm			10^5 g cm^{-1} s^{-2}	10^7 ergs
	10^9 nm				0.239 cal
	10^{10} Å				6.242x10^{18} eV
	10^{12} pm				1 C^2 Ω s^{-1}
					1V C

Pressure	Viscosity	Electric dipole moment	Charge	Current	Temperature
1Pa	1 g cm^{-1} s^{-1}	1 C m	1 C	1 A	K = ^{0}C + 273.15
1 N m^{-2}	1 dyne s cm^{-2}	3.00x10^{29} D	2.998x10^9 esu	1 C s^{-1}	^{0}C = 5/9(^{0}F − 32)
10^{-5} bar	1 poise				
9.871x10^{-6} atm					
7.502x10^{-3} mmHg					
7.502x10^{-3} torr					

Conversion Factors Between c.g.s. and m.k.s.a. Units:

1C = 2.998x10^9 esu or statC

1erg = 1 esu^2 cm^{-1}

1 D = 10^{-18} esu cm

1 esu cm^{-1} = 1 statV

1 A = 2.998x10^9 statA

1 V = 0.333x10^{-2} statV

1 Ω = 0.111x10^{-11} stat Ω

1 F = 9x10^{11} cm

Useful Constants

Symbol	Name	Value
c_0	speed of light	2.998×10^8 m s^{-1}
e_0	electron charge	1.602×10^{-19} C
F	Faraday's constant $= e_0 N_A$	9.649×10^4 C mol^{-1}
h	Plank's constant	6.626×10^{-34} J s
k	Boltzman's constant	1.381×10^{-23} J K^{-1}
N_A	Avogadro's number	6.022×10^{23} mol^{-1}
R	gas constant	8.314 J mol^{-1} K^{-1}
ε_0	permittivity of free space	8.854×10^{-12} C^{2}N^{-1}m^{-2}
$4\pi\varepsilon_0$		1.112×10^{-10} C^2 J^{-1} m^{-1}
0 K	absolute zero of temperature	-273.15 °C
π	pi	$3.14159...$

Other Data

Dielectric constant	$20\,^{\circ}C$	$25\,^{\circ}C$	$30\,^{\circ}C$	
Water	80.1	78.3	76.54	Dipole moment of water: $\mu_w = 1.87$ D
$\dfrac{\partial \varepsilon_w}{\partial T} = -0.356$ K^{-1}				Quadruple moment of water: $p_w = 3.9 \times 10^{-8}$ D cm
Cyclohexane	2.02	2.02		
Carbon tetrachloride	2.24	2.23		Polarizability of water: $\alpha_w = 1.46 \times 10^{-24}$ cm^3
Benzene	2.28	2.27		
Ammonia	17.35	16.9		
1-Propanol		20.1		
Ethanol		24.30		
Methanol	33.62	33.63		
Nitrobenzene	35.74	34.82		

Masses and Radii

H	Be	B	C	N	O	F
1.008	9.012	10.811	12.011	14.007	15.999	18.998
	(+2)35	(+3)23	(-4)260	(-3)171	(-2)132	(-)136
			(+4)160	(+3)16		
				(+5)13		

Li	Mg	Al	Si	P	S	Cl
6.941	24.305	26.982	28.086	30.974	32.062	35.453
(+) 59	(+2)66	(+3)51	(-4)2.71	(-3)212	(-2)184	(-)181
			(+4)0.42	(+3)44		
				(+5)35		

Na	Ca					Br
22.990	40.08					79.904
(+) 95	(+2)99					(-)195

K	Sr					I
39.098	87.62					126.905
(+)133	(+2)112					(-)216

Rb	Ba
85.468	137.3
(+)149	(+2)134

Cs
132.905
(+)167

mass → H_2O ← radius
18.015
138 pm

CHAPTER 2

ION-SOLVENT INTERACTIONS

EXERCISES

Review of Sections 2.1 to 2.4 of the Textbook.

How is an *electrochemical system* constituted? Describe the two ways mobile ions can create an interphase. What is the difference between *hydration* and *solvation*? Briefly, describe the three main approaches to the study of solvation. Draw a schematic of the water molecule. What is *libration*? Define a *dipole* and its *moment*. Explain how the water molecule is considered a dipole. What is the role of free orbitals in the water molecule? What is a *hydrogen bond*? Describe the ice and liquid water structures. Mention the seminal contributions of Bernal and Fowler, as well as of Frank and Wen, to the understanding of water structure. Write the basic equation for the interaction energy between particles and describe each one of its terms. How does an ion affect the structure of water? Draw a schematic showing the *primary*, the *structure-broken* and the *bulk* regions of water around an ion. What are the radius and the dipole moment of water? What energy changes are described in the Bernal-Fowler theory of ion-solvent interaction?

2.1 A positive charge $+e_0$, and a negative charge $-e_0$ are separated by 53 pm. Calculate their dipole moment. (Cf. Exercise 2.29 in the textbook) (Contractor)

Answer:

The dipole moment is given by:

$$\mu = \left(\text{charge}\right)\left(\text{separation of charges}\right)$$

$$= \left(1.602 \times 10^{-19}\, C\right)\left(53 \times 10^{-12}\, m\right) \times \frac{1\,D}{3.336 \times 10^{-30}\, C\,m} = 2.54\,D \quad (2.1)$$

Comment. According to Bohr's model, in the hydrogen atom the proton and the electron are separated 0.53 Å in the ground state. They form a dipole with a moment of 2.54 D.

2.2. **Water often is subjected to extremely strong electric fields as, for example, in the solvation of ions. By considering the total H-bond energy of water, calculate the electric-field strength that will break up H-bonded water. (cf. Exercise 2.33 in the textbook) (Bockris-GamboaAldeco)**

Data:

From Section 2.4 in the textbook: $10\,kJ\,mol^{-1} < W_{\text{H-bonding}} < 40\,kJ\,mol^{-1}$

Answer:

The relationship between energy and electric field is:

$$W_{a-b} = q\vec{X}d \quad (2.2)$$

Considering q to be the charge of the proton, and the average hydrogen-bonding energy as 25 kJ mol^{-1}, the electric field strength needed to break the hydrogen bond is

$$\vec{X} = \frac{W_{a-b}}{e_0 d} = \frac{25000\,J\,mol^{-1}}{\left(1.602 \times 10^{-19}\, C\right)\left(1m\right)} \times \frac{1\,mol}{6.022 \times 10^{23}} \times \frac{1\,V\,C}{1\,J} \quad (2.3)$$

$$= 0.260\ V\,m^{-1}$$

2.3 (a) Show that the electrostatic interaction potential energy between two charges q_1 and q_2 can be represented in the non-conventional form, although widely used equation,

$$U = \frac{332}{\varepsilon} \frac{z_1 z_2}{r_{1-2}}$$ (2.4)

where U is the potential energy given in units of kcal mol^{-1}, r_{1-2} is the distance between the charges expressed in angstroms (Å), ε is the relative dielectric constant of the medium (unitless), and z_1 and z_2 are given by the equation $q_i = z_i e_0$ where q_i represent the charge expressed in electronic charge units (esu). **(b) What is the Coulombic interaction energy in kcal mol^{-1} between a proton and an electron separated by 1 Å in vacuum? (cf. Exercise 2.27 in the textbook) (Contractor-Casidar)**

Answer:

(a) The corresponding potential-energy equation given in SI units is:

$$U = \frac{1}{4\pi\varepsilon_0 \varepsilon} \frac{q_1 q_2}{r_{1-2}}$$ (2.5)

where q_1 and q_2 are expressed in coulombs (C), r_{1-2} in meters (m), ε is unitless, ε_0, the permittivity of free space, is a constant equal to 8.854×10^{-12} C^{2}J^{-1}m^{-1}, and U is the potential energy given in units of J. However, this expression can be written in *cgs* units as (cf. Chapter 1)

$$U = \frac{q_1 q_2}{\varepsilon r_{1-2}} = \frac{z_1 z_2 e_0^2}{\varepsilon r_{1-2}}$$ (2.6)

where U is given in ergs, the electron charge, e_0, is given in esu or electrostatic units and r_{1-2} in cm. With r_{1-2} given in Å, substituting the value of e_0, multiplying by Avogadro's number, N_A, and applying some conversion factors gives,

$$U = \frac{z_1 z_2}{\varepsilon r_{1-2} [\text{Å}]} \left(4.8 \times 10^{-10} \text{ esu} \right)^2 \left(6.022 \times 10^{23} \text{ mol}^{-1} \right)$$ (2.7)

$$\times \frac{10^8 \text{ Å}}{1 \text{cm}} \times \frac{1 \text{erg}}{1 \text{esu}^2 \text{cm}^{-1}} \times \frac{1 \text{J}}{10^7 \text{ erg}} \times \frac{1 \text{cal}}{4.186 \text{J}} \times \frac{1 \text{kcal}}{1000 \text{cal}}$$

or

$$U\left[\text{kcal mol}^{-1} \right] = \frac{332 \; z_1 z_2}{\varepsilon \, r_{1-2} \left[\text{Å} \right]} \tag{2.8}$$

(b) When $r_{1-2} = 1$ Å and $\varepsilon = 1$, then,

$$U\left[\text{kcal mol}^{-1} \right] = \frac{332 \; (1)(-1)}{(1)(1 \text{Å})} = -332 \text{ kcal mol}^{-1} \tag{2.9}$$

Review of Section 2.5 of the Textbook

What are the *thermodynamic, transport,* and *spectroscopic* methods used to study solvation? Define *heats of solvation, sublimation* and *dissolution*. Write an equation for the change of chemical potential of dissolution. Describe a method to determine $\Delta G^0_{solvation}$, $\Delta S^0_{solvation}$, and $\Delta H^0_{solvation}$. What is the importance of these quantities?

2.4 Calculate the sum of the heats of hydration of K^+ and F^- ions. The lattice energy is –194.7 kcal mol^{-1}. The heat of the solution is – 4.1 kcal mol^{-1}. (Cf. Exercise 2.11 in the textbook) (Bockris-GamboaAldeco)

Answer:

The heat of solvation of a salt is related to the heat of solution and the heat of sublimation by the following equation (cf. Eq. 2.3 in the textbook):

$$\Delta H_{s,salt} = \Delta H_{soln} - L_{sub} \tag{2.10}$$

However, the heat of sublimation is equal to the lattice energy but has an opposite sign. Therefore,

$$\Delta H_{s,salt} = \Delta H_{soln} + \Delta H_{lattice} \tag{2.11}$$

Substituting the corresponding values,

$$\Delta H_{s,salt} = -4.1 \, \text{kcal mol}^{-1} - 194.7 \, \text{kcal mol}^{-1} = -198.8 \, \text{kcal mol}^{-1}$$

$$= -832.2 \, \text{kJ mol}^{-1} \tag{2.12}$$

Review of Sections 2.6 and 2.7 of the Textbook.

Why are measurements of *partial-molar volumes* important in understanding solvation? Explain Conway's method to determine ionic volumes of individual ions. Write an expression relating the *compressibility* of the solution to solvation. Explain how solvation numbers are obtained from Passynski's method. What important finding did Orani contribute to Passynski's theory? How is the *compressibility* of a liquid measured?

2.5 The adiabatic compressibilities of water and of a 0.1 M NaI solution at 298 K are 4.524×10^{-10} Pa^{-1} and 4.428×10^{-10} Pa^{-1}, respectively. Calculate the hydration number of the NaI molecule when the density of the solution is 1.0086 g cm^{-3}. (Cf. Exercise 2.18 in the textbook) (Kim)

Data:

$\beta_0 = 4.524 \times 10^{-10}$ Pa^{-1} $\rho_{soln} = 1.0086$ g cm^{-3} $T = 298$ K
$\beta = 4.428 \times 10^{-10}$ Pa^{-1} $c = 0.1$ mol dm^{-3}

Answer:

The hydration number of a salt can be obtained from (c.f. Eq. 2.11 in the textbook):

$$n_s = \frac{n_w}{n_{NaI}} \left(1 - \frac{\beta}{\beta_0} \right) \tag{2.13}$$

where n_w is the number of water moles and n_{NaI} the number of electrolyte moles.

The value of n_{NaI} is 0.1 mol in 1000 cm^3 of solution. The value of n_w, in 1000 cm^3 of solution can be calculated as

$$n_w = \frac{\left(V_{soln} \rho_{soln} - n_{NaI} (MW)_{NaI} \right)}{(MW)_w} \tag{2.14}$$

$$= \left[\left(1000\,cm^3 \right) \left(1.0086\,g\,cm^{-3} \right) - \left(0.1\,mol \right) \left(148.9\,g\,mol^{-1} \right) \right] \frac{1\,mol}{18.02\,g}$$

$$= 55.14\,mol$$

Therefore, the hydration number of NaI is,

$$n_s = \frac{n_w}{n_{NaI}} \left(1 - \frac{\beta}{\beta_0} \right) = \frac{55.14\,mol}{0.1\,mol} \left(1 - \frac{4.428 \times 10^{-10}\,Pa^{-1}}{4.524 \times 10^{-10}\,Pa^{-1}} \right) = 11.7 \qquad (2.15)$$

2.6 The adiabatic compressibility of water at 25 °C is 4.524x10⁻¹⁰ Pa⁻¹. Calculate the adiabatic compressibility of 0.101 M solution of CaCl₂ of density 1.0059 g cm⁻³ at 25 °C, if the hydration number of the electrolyte is 12. (Cf. Exercise 2.19 in the textbook) (Contractor)

Data:

$\beta_0 \quad = 4.524 \times 10^{-10}\,Pa^{-1}$ $\qquad\qquad\qquad \rho_{soln} = 1.0059\,g\,cm^{-3}$

$c_{CaCl2} = 0.101\,mol\,dm^{-3}$ $\qquad\qquad\qquad\qquad n_s \quad = 12$

Answer:

The hydration number in terms of compressibility data can be calculated from (cf. Eq. 2.11 in textbook):

$$n_s = \frac{\alpha\,n_w}{n_{CaCl_2}} = \frac{n_w}{n_{CaCl_2}} \left(1 - \frac{\beta}{\beta_0} \right) \qquad (2.16)$$

where n_w is the number of water moles and n_{CaCl2} the number of electrolyte moles. The value of n_w is calculated as

$$n_w = \frac{\rho_{soln} - n_{NaCl_2}\,(MW)_{NaCl_2}}{(MW)_w}$$

$$= \frac{1.0059\,g\,dm^{-3} - \left(0.101\,mol\,dm^{-3} \right) \left(111\,g\,mol^{-1} \right)}{18.016\,g\,mol^{-1}} \qquad (2.17)$$

$$= 55.21\,mol\,in\,1\,dm^3\,of\,solution$$

Therefore,

$$\beta = \beta_0 \left(1 - \frac{n_{CaCl_2} n_s}{n_w} \right) = 4.524 \times 10^{-10} \, Pa \left[1 - \frac{(0.101 \, mol)(12)}{55.21 \, mol} \right] \quad (2.18)$$

$$= 4.425 \times 10^{-10} \, Pa$$

2.7 In the textbook are data on compressibility as a function of concentration. Use the Passynski equation to calculate the total salvation number of NaBr at infinite dilution. (cf. Exercise 2.21 in the textbook) (Bockris-GamboaAldeco)

Data:

From Table 2.5 in the textbook, the compressibility of NaBr aqueous solutions at different concentrations:

c (M)	5.2	2.0	1.1	0.1	0.05
$\beta \times 10^6$ (bar^{-1})	26.93	35.99	39.75	44.23	44.47

$\beta_w = 45.24 \times 10^{-6}$ bar^{-1} at 25 ^{0}C

Answer:

The Passynski equation reads (cf. Eq. 2.11 in the textbook):

$$n_s = \alpha \frac{n_1}{n_2} = \left(1 - \frac{\beta}{\beta_0} \right) \frac{n_1}{n_2} \quad (2.19)$$

where β_0 is the compressibility of the solvent (water), and β the compressibility of the solution; n_1 is the number of solvent (water) moles and n_2 the number of salt moles. In 1 liter of solution there are approximately 55 mol of water (n_w).

For a solution of 5.2 M NaBr,

$$n_s = \left(1 - \frac{26.93 \times 10^{-6} \, bar^{-1}}{45.24 \times 10^{-6} \, bar^{-1}} \right) \frac{55 mol}{5.2 mol} = 4.28 \quad (2.20)$$

In the same way for the other solutions,

c (M)	5.2	2.0	1.1	0.1	0.05
n_s	4.28	5.62	6.07	12.28	18.7

Plotting these values as in Fig. 2.1 and neglecting the last two points, gives a straight line of equation,

$$n_s = 6.52 - 0.43c \qquad (2.21)$$

Therefore, when $c \to 0$ the value of the total solvation number of NaBr at infinite dilution can be obtained. This number is, from the graph, $n_s^\infty = 6.52$

2.8 Sound velocity in water is measured to be 1496.95 m s^{-1} at 25 ^{0}C. Calculate the adiabatic compressibility of water in bar^{-1}. (Cf. Exercise 2.20 in the textbook) (Xu)

Data:

$c_0 = 1496.95$ m s^{-1} $\rho_w = 1$ g cm$^{-3} = 10^3$ kg m^{-3} $T = 298$ K

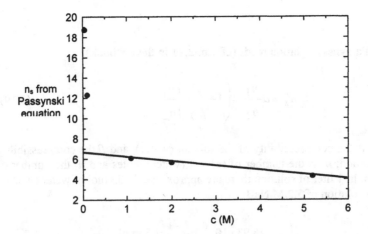

Figure 2.1. Total solvation number of NaBr at different concentrations according to Passynski's equation.

Answer:

The relationship between the compressibility of a liquid or solution and the velocity of sound in it is (c.f. Eq. 2.12 in the textbook),

$$\beta = \frac{1}{c_0^2 \, \rho_w} \tag{2.22}$$

Substituting values in Eq. (2.22),

$$\beta = \frac{1}{\left(1496.95\,\text{m s}^{-1}\right)^2 \left(10^3 \, \text{kg m}^{-3}\right)} \times \frac{1\,\text{kg m s}^{-2}}{1\,\text{N}} \times \frac{1\,\text{N m}^{-2}}{1\,\text{Pa}} \times \frac{10^5 \, \text{Pa}}{1\,\text{bar}}$$

$$= 4.46 \times 10^{-5} \, \text{bar}^{-1} \tag{2.23}$$

Review of Sections 2.8 to 2.11 of the Textbook

Describe the *ultrasound method* for the determination of hydration numbers. Why does the *mass-action law* of chemistry not work for ions in solution? What is an *activity coefficient*? How does the activity coefficient vary with concentration? Explain how this variation can be used to determine hydration numbers. Mention some drawbacks of this method. Describe briefly how the determination of the *mobility* of ions can lead to information of solvation numbers. Mention some difficulties found within this method. How are spectroscopy techniques used to obtain hydration numbers? What regions of the infrared spectrum give information about *intermolecular* and *intramolecular* effects? What information in the infrared region can be obtained related to the structure of water and its interaction with ions? Write Brag's equation for X-rays analysis. Explain the difficulties found in attempting to utilize the powerful tool of neutron diffraction to the study of solvation. What is a *distribution function*? What is the relation of *coordination waters* and *hydration waters*? How are distribution functions applied to determine coordination and hydration numbers? How are these quantities determined? What is the contribution of Hewich, Neilson and Enderby to the structure of ions and their solvation sheets? Describe possible nonelastic-scattering effects in Raman spectroscopy. Mention some important findings related to solution structure by Raman spectroscopy.

2.9 Calculate the hydration number of Na$^+$ when the mobility of the ion in water is 44×10^{-5} cm^2 s^{-1} V^{-1}, and the viscosity of the solution is 0.01 poise.

The Stokes radius is given by $r_{Stokes} = z_i e_0 / 6\pi\eta u$. (Cf. Exercise 2.23 in the textbook) (Kim)

Data:

$r_{Na^+} = 95$ pm $\qquad\qquad r_w = 138$ pm $\qquad\qquad u_{Na^+} = 44 \times 10^{-5}$ cm^2 s^{-1} V^{-1}

$\eta_{soln} = 0.01$ poise

$$= 0.01 \frac{dyn\ s}{cm^2} \times \frac{1N}{10^5\ dyn} \times \frac{1m}{1m} \times \frac{1J}{1Nm} \times \frac{1m}{100cm} = 10^{-9}\ J\ s\ cm^{-3}$$

Answer:

The hydration number of the ion can be obtained from (cf. Eq. 2.23 in textbook):

$$n_s = \frac{r_{Stokes}^3 - r_{crys}^3}{r_w^3} \tag{2.24}$$

The radius of the hydrated ion is obtained from the Stokes equation given above (cf. Eq. 4.183 in textbook):

$$r_{Stokes} = \frac{z_i e_0}{6\pi\eta u} = \frac{(+1)\left(1.602 \times 10^{-19}\ C\right)}{6\pi\left(10^{-9}\ J\ s\ cm^{-3}\right)\left(44 \times 10^{-5}\ cm^2\ s^{-1}\ V^{-1}\right)} \times \frac{1J}{1CV} \tag{2.25}$$

$$= 1.93 \times 10^{-8}\ cm$$

Therefore, Eq. (2.24) gives

$$n_s = \frac{r_{Stokes}^3 - r_{crys}^3}{r_w^3} = \frac{(193\,pm)^3 - (95\,pm)^3}{(138\,pm)^3} = 2.4 \tag{2.26}$$

2.10 An IR spectrum has a peak with a wavenumber of 1.561×10^3 cm^{-1}. (a) What is the wavelength and frequency of the corresponding bond? (b) If the spectrum originates from water, calculate the force constant between O and H. (cf. Exercise 2.34 in the textbook) (Bockris-GamboaAldeco)

Answer:

(a) The wavelength (λ) and frequency (v) are related to the wavenumber ($\tilde{v}$) by $\lambda = \dfrac{1}{\tilde{v}}$ and $v = c_0 \tilde{v}$. Substituting the corresponding values gives:

$$\lambda = \frac{1}{\tilde{v}} = \frac{1}{3.561 \times 10^3 \text{ cm}^{-1}} = 2.808 \times 10^{-4} \text{ cm} \tag{2.27}$$

and

$$v = c_0 \tilde{v} = \left(2.998 \times 10^{10} \text{ cm s}^{-1} \right) \left(3.561 \times 10^3 \text{ cm}^{-1} \right)$$
$$= 1.068 \times 10^{14} \text{ s}^{-1} \tag{2.28}$$

(b) The frequency (v) and the force constant (k) are related through the equation (cf. Eq. 2.24 in textbook) $v = \dfrac{1}{2\pi} \sqrt{\dfrac{k}{\tilde{\mu}}}$, where $\tilde{\mu}$ is the reduced mass, given by $\dfrac{1}{\tilde{\mu}} = \dfrac{1}{m_1} + \dfrac{1}{m_2}$. Considering that the subscript 1 refers to the oxygen of the water molecule, and the subscript 2 to one of its hydrogens, then, the reduced mass of water is given by

$$\frac{1}{\tilde{\mu}} = \frac{1}{16} + \frac{1}{1} = \frac{1}{0.94 \text{ g mol}^{-1}} \tag{2.29}$$

or

$$\tilde{\mu} = 0.94 \text{ g mol}^{-1} \frac{1 \text{ mol}}{6.022 \times 10^{23}} = 1.56 \times 10^{-24} \text{ g} \tag{2.30}$$

Then, the force constant is,

$$k = (2\pi v)^2 \tilde{\mu} = 4\pi^2 \left(1.068 \times 10^{14} \text{ s}^{-1} \right)^2 \left(1.56 \times 10^{-27} \text{ Kg} \right) \times \frac{1 \text{N}}{1 \text{Kg m s}^{-2}}$$
$$= 702.47 \text{ N m}^{-1} \tag{2.31}$$

Review of Sections 2.12 to 2.14 of the Textbook

Define *dielectric constant*. Why does the dielectric constant decrease when ions are added to a liquid? How does the dielectric constant vary as a function of the distance from an ion? Write the equation proposed by Hasted relating dielectric constants to solvation numbers. What is *electrostriction*? What differences exist between cations and anions on the frequency at which water undergoes relaxation? How are the dielectric constants of liquids and of ionic solutions measured? How can these measurements be used to study the structure of ionic solutions? Explain a method to determine the hydration energy of the outer-hydration shell. How is this value relative to the total hydration energy? What parameters determine the solvation number of ions? Mention an approach to determine *ionic* solvation numbers.

2.11 If the primary hydration number of NaCl in 1 M solution is six, make a rough calculation of the dielectric constant of the solution if the dielectric constant of pure water is taken as 80. (cf. Exercise 2.6 in the textbook) (Bockris-GamboaAldeco)

Answer:

An equation to calculate the dielectric constant of a solution is (cf. Eq. 2.27 in the textbook):

$$\varepsilon_{soln} = 80\left(\frac{55 - c_i n_s}{55}\right) + 6\left(\frac{c_i n_s}{55}\right) \tag{2.32}$$

Substituting the corresponding parameters in Eq. (2.32),

$$\varepsilon_{soln} = 80\left[\frac{55\,\text{mol dm}^{-3} - \left(1\,\text{mol dm}^{-3}\right)(6)}{55\,\text{mol dm}^{-3}}\right] + 6\left[\frac{\left(1\,\text{mol dm}^{-3}\right)(6)}{55\,\text{mol dm}^{-3}}\right]$$

$$= 71.9 \tag{2.33}$$

2.12 Solvation numbers for Na$^+$ and Cl$^-$ have been measured as 5 and 1, respectively. With this information, and recalling that the number of moles per liter of pure water is 55, calculate the dielectric constant of a 5 M

solution of NaCl. The dielectric constant of pure water is to be taken as 80 near room temperature. When the water molecules are held immobile in respect to the variations of an applied field, ε drops to 6. (Cf. Exercise 2.10 in the textbook) (Bockris-GamboaAldeco)

Data:

$$n_{s,Na^+} = 5 \qquad\qquad n_w = 55\,\text{mol in 1 liter} \qquad\qquad \varepsilon_{w,bulk} = 80$$

$$n_{s,Cl^-} = 1 \qquad\qquad c_i = 5\,\text{M} \qquad\qquad \varepsilon_{w,saturated} = 6$$

Answer:

According to Hasted, a simple equation to determine the dielectric constant of a solution is [cf. Section 2.12.1 in the textbook]:

$$\varepsilon_{soln} = \varepsilon_{w,bulk}\,\frac{n_{w,bulk}}{n_w} + \varepsilon_{w,saturated}\,\frac{n_{w,saturated}}{n_w} \qquad (2.34)$$

The number of saturated water molecules can be taken as the number of water molecules around the ions. In 1 liter of solution,

$$n_{w,saturated} = c_{Na^+}\,n_{s,Na^+} + c_{Cl^-}\,n_{s,Cl^-}$$
$$= (5\,\text{mol})(5) + (5\,\text{mol})(1) = 30\,\text{mol} \qquad (2.35)$$

and the number of water molecules in the bulk of the solution is:

$$n_{w,bulk} = n_w - n_{w,saturated} = 55\,\text{mol} - 30\,\text{mol} = 25\,\text{mol} \qquad (2.36)$$

Therefore,

$$\varepsilon_{soln} = (80)\frac{25\,\text{mol}}{55\,\text{mol}} + 6\frac{30\,\text{mol}}{55\,\text{mol}} = 39.6 \qquad (2.37)$$

Review of Sections 2.15.1 to 2.15.11 of the Textbook

Describe the Halliwell-Nyburg method to determine relative heats of hydration of individual ions. Do opposite charged ions of equal radii have equal heats of solvation? Draw a schematic of water as a *quadrupole*. Which are the steps involved in the ion-quadrupole model of ion-solvent interactions? Write expressions for (a) the ion-dipole energy in the gas phase, (b) the energy involved in the introduction of solvated ions into a cavity (Born term), and (c) the full energy equations for cation- and anion-solvent interactions. What is *deformation polarizability*? How are the equations for heat of ion-solvent interactions modified by the *induced-dipole* effect? Explain a method to determine the individual heats of hydration from values known for the salt. How did Lee and Tai determine the absolute standard entropy of the proton in solution? Draw schematics of the solvation model of Bockris and Reddy. What is understood by *dynamic solvation number*? Explain the steps involved in the cycle of ion-solvent interactions proposed by Bockris and Saluja. Write Born's equation for solution energy. What is the meaning of each one of its terms? Which are the difficulties presented by the Born's theory of solvation? What does the acronyms *SN*, *CN* and *NSCW* represent? Write Bockris and Saluja's equations for the interaction energies (a) of an ion with waters oriented along the ionic field, (b) of an ion with nonsolvationally coordinated waters, and (c) between lateral ions. What are the differences among the structure-breaking energies proposed by (a) Bernal and Fowler, (b) Frank and Wen, and (c) Bockris and Reddy? How do these models fit experimental data?

2.13 Calculate the absolute heats of hydration of Na^+ and Cl^- using the absolute heat of hydration of H^+ of -1113.0 kJ mol^{-1}. The heat of interaction between HCl and water is -1454.0 kJ mol^{-1}, the heat of solution of NaCl is $+3.8$ kJ mol^{-1}, and the heat of sublimation of NaCl is $+772.8$ kJ mol^{-1}. (Cf. Exercise 2.17 in the textbook) (Kim)

Data:

$$\Delta H_{HCl}\ (hyd) = -1454.0\ \text{kJ mol}^{-1} \qquad \Delta H_{H^+}\ (hyd) = -1113.0\ \text{kJ mol}^{-1}$$

$$\Delta H_{NaCl}\ (so\ln) = +3.8\ \text{kJ mol}^{-1} \qquad \Delta H_{NaCl}\ (sub) = +772.8\ \text{kJ mol}^{-1}$$

Answer:

The heat of hydration of HCl is (cf. Eq. 2.29 in the textbook):

$$\Delta H_{HCl}\ (hyd) = \Delta H_{H^+}\ (hyd) + \Delta H_{Cl^-}\ (hyd) = -1454.0\ \text{kJ mol}^{-1} \quad (2.38)$$

From Eq. (2.38),

$$\Delta H_{Cl^-} (hyd) = -1454.0 \text{kJ mol}^{-1} - \left(-113.0 \text{kJ mol}^{-1} \right)$$
$$= -341.15 \text{kJ mol}^{-1}$$

(2.39)

In the same way, $\Delta H_{NaCl}(hyd)$ is also equal to:

$$\Delta H_{NaCl} (hyd) = \Delta H_{Na^+} (hyd) + \Delta H_{Cl^-} (hyd)$$

(2.40)

However, the heat of solvation of NaCl is also given by [cf. Fig.2.13 in textbook],

$$\Delta H_{NaCl} (hyd) = \Delta H_{NaCl} (so\ln) - \Delta H_{NaCl} (sub)$$
$$= +3.8 \text{kJ mol}^{-1} - 772.8 \text{kJ mol}^{-1} = -769.0 \text{kJmol}^{-1}$$

(2.41)

Therefore, the absolute heat of hydration of Na^+ ions is

$$\Delta H_{Na^+} (hyd) = \Delta H_{NaCl} (hyd) - \Delta H_{Cl^-} (hyd) = -769.0 \text{ kJ mol}^{-1}$$
$$- \left(-341.1 \text{kJ mol}^{-1} \right) = -427.9 \text{ kJ mol}^{-1}$$

(2.42)

2.14 The Cl⁻ ion has a radius of 181 pm. Find how much larger the ion-dipole term is compared with the Born term. The dielectric constant in the Born equation is to be taken as 80. Neglect any dependence of the dielectric constant with temperature. (cf. Exercise 2.2 in the textbook) (Bockris-GamboaAldeco)

Answer:

The ion-dipole interaction energy in the gas phase is given by (cf. Eq. 2.41 in the textbook):

$$W_{I-D} = -\frac{4 N_A z_i e_0 \mu_w}{4 \pi \varepsilon_0 (r_i + r_w)^2}$$

(2.43)

where the term $4\pi\varepsilon_0$ in the denominator has been introduced to write the equation in the *mksa* system, i.e., the charge in Coulombs, the distance in meters, and the free energy in joules per mol. For Cl⁻,

$$W_{I-D} = -\frac{4\left(6.022 \times 10^{23} \text{ mol}^{-1}\right)|-1|\left(1.602 \times 10^{-19} \text{ C}\right)(1.87 \text{ D})}{\left(1.112 \times 10^{-10} \text{ C}^2 \text{J}^{-1} \text{m}^{-1}\right)(181 \text{ pm} + 138 \text{ pm})^2} \qquad (2.44)$$

$$\times \frac{3.336 \times 10^{-30} \text{ C m}}{1 \text{ D}} \times \frac{10^{24} \text{ pm}^2}{1 \text{ m}^2} = -212.8 \text{ kJ mol}^{-1}$$

The Born term is given in the *mksa* system by (cf. Eq. 2.40 in the textbook):

$$W_{BC} = -\frac{N_A \left(z_i e_0\right)^2}{2\left(r_i + 2r_w\right) 4\pi\varepsilon_0}\left(1 - \frac{1}{\varepsilon} - \frac{T}{\varepsilon^2}\frac{\partial\varepsilon}{\partial T}\right) \qquad (2.45)$$

Substituting the corresponding parameters for Cl⁻ and taking into account that $\dfrac{\partial\varepsilon}{\partial T} = 0$,

$$W_{BC} = -\frac{\left(6.022 \times 10^{23} \text{ mol}^{-1}\right)(-1)^2\left(1.602 \times 10^{-19} \text{ C}\right)^2}{2\left(1.112 \times 10^{-10} \text{ C}^2 \text{J}^{-1} \text{m}^{-1}\right)[181 \text{ pm} + 2(138 \text{ pm})]}\left(1 - \frac{1}{80}\right) \quad (2.46)$$

$$\times \frac{10^{12} \text{ pm}}{1 \text{ m}} = -150.2 \text{ kJ mol}^{-1}$$

Therefore,

$$\frac{W_{I-D} - W_{BC}}{W_{BC}} = \frac{-212.8 \text{ kJ mol}^{-1} + 150.2 \text{ kJ mol}^{-1}}{-150.2 \text{ kJ mol}^{-1}} = 0.417 \quad (2.47)$$

That is, the ion-dipole term is **41.7%** larger than the Born term.

2.15 The interaction energy of the iodide ion with water dipoles in gas phase is –171 kJ mol⁻¹ at 298 K. Calculate the radius of the water molecule. Take the iodide radius and the dipole moment of water from tables in Chapter 1. (Cf. Exercise 2.3 in the textbook) (Contractor)

Data:

$$r_{I^-} = 216 \text{ pm} \qquad\qquad \mu_w = 1.87 \text{ D} \qquad\qquad W_{I-D} = -171 \text{ kJ mol}^{-1}$$

Answer:

The ion-dipole interaction in the gas phase is given in the *mksa* system by (c.f. Eq. 2.41 in textbook):

$$W_{I-D} = -\frac{4 N_A z_i e_0 \mu_w}{4 \pi \varepsilon_0 (r_i + r_w)^2} \tag{2.48}$$

Solving for r_w and substituting the corresponding values,

$$r_w = \left(-\frac{4 N_A z_i e_0 \mu_w}{4 \pi \varepsilon_0 W_{I-D}} \right)^{1/2} - r_i \tag{2.49}$$

$$= \left[-\frac{4 \left(6.022 \times 10^{23} \, mol^{-1} \right) |-1| \left(1.602 \times 10^{-19} \, C \right) (1.87 D)}{\left(1.112 \times 10^{-10} \, C^2 J^{-1} m^{-1} \right) \left(-171000 \, J \, mol^{-1} \right)} \right]$$

$$\times \frac{3.336 \times 10^{-30} \, Cm}{1 D} \Big]^{1/2} - 2.16 \times 10^{-10} \, m = 1.398 \times 10^{-10} \, m$$

$$= 140 \text{ pm}$$

2.16 The heat of interaction of chloride ion with water is –347.3 kJ mol $^{-1}$, and the corresponding contribution to the Born charging process is –152.5 kJ mol $^{-1}$. (a) If the dielectric constant of water at 298 K is 78.3, estimate its rate of change with temperature at this temperature. (b) Calculate the percent error introduced in the Born charging term if the dielectric constant is assumed independent of temperature. (Cf. Exercise 2.4 in the textbook) (Contractor)

Data:

$$\Delta H_{Cl^- - w} = -347.3 \, kJ \, mol^{-1} \qquad r_{Cl^-} = 181 \, pm \qquad T = 298 \, K$$

$$W_{BC} = -152.5 \, kJ \, mol^{-1} \qquad r_w = 138 \, pm \qquad \varepsilon_w = 78.3$$

Answer:

(a) The interaction energy involved in the process of introducing the primary solvated ion formed in the gas phase into the solvent cavity, i.e., the Born term, is given in the *mksa* system by (c.f. Eq. 2.42 in textbook):

$$W_{BC} = -\frac{N_A (z_i e_0)^2}{2(r_i + 2r_w)4\pi\varepsilon_0}\left(1 - \frac{1}{\varepsilon} - \frac{T}{\varepsilon^2}\frac{\partial\varepsilon}{\partial T}\right) \tag{2.50}$$

Solving for $\partial\varepsilon/\partial T$ and substituting the corresponding values,

$$\frac{\partial\varepsilon}{\partial T} = \frac{\varepsilon^2}{T}\left[\frac{2W_{BC}(r_i + 2r_w)4\pi\varepsilon_0}{N_A(z_i e_0)^2} + 1 - \frac{1}{\varepsilon}\right] \tag{2.51}$$

$$= \frac{78.5^2}{298\,K}\left[\frac{2\left(-152500\,J\,mol^{-1}\right)(181 + 276)\times10^{-12}\,m}{6.022\times10^{23}\,mol^{-1}\left(1.602\times10^{-19}\,C\right)^2}\right.$$

$$\left.\times\left(1.112\times10^{-10}\,C^2J^{-1}m^{-1}\right) + 1 - \frac{1}{78.5}\right] = -0.32$$

(b) If the dielectric constant is assumed independent of temperature, $\partial\varepsilon/\partial T = 0$, then,

$$W_{BC} = -\frac{N_A(z_i e_0)^2}{2(r_i + 2r_w)4\pi\varepsilon_0}\left(1 - \frac{1}{\varepsilon}\right)$$

$$= -\frac{\left(6.022\times10^{23}\,mol^{-1}\right)\left(1.602\times10^{-19}\,C\right)^2}{2(181 + 276)\times10^{-12}\,m\left(1.112\times10^{-10}\,C^2J^{-1}m^{-1}\right)} \tag{2.52}$$

$$\times\left(1 - \frac{1}{78.5}\right) = -150147\,J\,mol^{-1}$$

The error introduced in neglecting $\partial\varepsilon/\partial T$ is calculated as:

$$\frac{(W_{BC})_{\partial\varepsilon/\partial T} - W_{BC}}{W_{BC}} \times 100 = \frac{-150.2\,\text{kJ}\,\text{mol}^{-1} + 152.5\,\text{kJ}\,\text{mol}^{-1}}{152.5\,\text{kJ}\,\text{mol}^{-1}} \times 100 = -1.5\%$$

(2.53)

2.17 When fluoride ions encounter water molecules, the ion-quadrupole interaction energy is –394.6 kJ mol⁻¹. Calculate the quadrupole moment of water at 298 K. (Cf. Exercise 2.5 in the textbook) (Contractor)

Data:

$T = 298\,\text{K}$ $\mu_w = 1.87\,\text{D}$ $\varepsilon_w = 78.3$

$W_{F^- - Q} = -394.6\,\text{kJ}\,\text{mol}^{-1}$ $r_{F^-} = 136\,\text{pm}$ $r_w = 138\,\text{pm}$

Answer:

The ion-quadrupole interaction energy is given in the *mksa* system by (c.f. Eq. 2.46 in textbook):

$$W_{i-Q} = -\frac{4N_A z_i e_0 \mu_w}{(r_i + r_w)^2\, 4\pi\varepsilon_0} - \frac{4N_A z_i e_0 p_w}{2(r_i + r_w)^3\, 4\pi\varepsilon_0}$$

(2.54)

Solving for p_w and substituting the corresponding values,

$$p_w = -\frac{2(r_i + r_w)^3\, 4\pi\varepsilon_0}{4N_A z_i e_0}\left(W_{i-Q} + \frac{4N_A z_i e_0 \mu_w}{(r_i + r_w)^2\, 4\pi\varepsilon_0} \right)$$

$$= -\frac{2\left[(136 + 138)\times 10^{-12}\,\text{m}\right]^3 \left(1.112\times 10^{-10}\,\text{C}^2\,\text{J}^{-1}\,\text{m}^{-1}\right)}{4\left(6.022\times 10^{23}\,\text{mol}^{-1}\right)\left(1.602\times 10^{-19}\,\text{C}\right)}$$

(2.55)

$$\times \left\{ -394.7\,k\text{J}\,\text{mol}^{-1} + \frac{4\left(6.022\times 10^{23}\,\text{mol}^{-1}\right)}{\left[(136 + 138)\times 10^{-12}\,\text{m}\right]^2} \right.$$

$$\times \frac{\left(1.602 \times 10^{-19} \, C\right)\left(1.87 \, D\right)}{\left(1.112 \times 10^{-10} \, C^2 J^{-1} m^{-1}\right)} \times \frac{3.336 \times 10^{-30} \, C \, m}{1 \, D} \Bigg\} \times \frac{1 \, D}{3.336 \times 10^{-30} \, C \, m}$$

$$= 3.8 \times 10^{-10} \, D \, m$$

2.18 Evaluate the heat of solvation for the ions K^+, Ca^{+2}, F^-, and Cl^- on terms of the ion-dipole approach. Neglect the quadrupole effect of water. The variation of the dielectric constant of the solution with temperature is $0.356 \, K^{-1}$. (Cf. Exercise 2.15 in the textbook) (Kim)

Data:

$$r_{K^+} = 133 \, pm \qquad\qquad r_{F^-} = 136 \, pm \qquad\qquad \varepsilon_w (298 \, K) = 78.3$$

$$r_{Ca^{+2}} = 99 \, pm \qquad\qquad r_{Cl^-} = 181 \, pm \qquad\qquad \frac{\partial \varepsilon}{\partial T} = 0.356 \, K^{-1}$$

Answer:

The heat of solvation for cations expressed in the *mksa* system is (cf. Eq. 2.47 in the textbook),

$$\Delta H_{I-S} = 20 - \frac{4 N_A z_i e_o \mu_W}{(r_i + r_W)^2 \, 4\pi\varepsilon_0}$$
$$- \frac{N_A (z_i e_o)^2}{2(r_i + 2r_W) 4\pi\varepsilon_0} \left[1 - \frac{1}{\varepsilon} - \frac{T}{\varepsilon^2} \left(\frac{\partial \varepsilon}{\partial T} \right)_P \right] \qquad (2.56)$$

Therefore, the enthalpy change for the solvation of K^+ in water is

$$\Delta H_{K^+ - w} = 20 \, kcal \, mol^{-1} \times \frac{4.186 \, kJ}{1 \, kcal} - \left\{ \frac{4 \times 6.023 \times 10^{23} \, mol^{-1} \, (+1)}{\left(133 \times 10^{-12} \, m + 138 \times 10^{-12} \, m \right)^2} \right.$$

$$\times \frac{\left(1.602 \times 10^{-19}\,C\right)1.8\,D}{\left(1.112 \times 10^{-10}\,C^2 J^{-1} m^{-1}\right)} \times \frac{3.336 \times 10^{-30}\,Cm}{1\,D} \Bigg\} \quad (2.57)$$

$$-\Bigg\{ \frac{6.023 \times 10^{23}\,mol^{-1}\,(+1)^2 \left(1.602 \times 10^{-19}\,C\right)^2}{2\left(133 \times 10^{-12}\,m + 138 \times 10^{-12}\,m\right)\left(1.112 \times 10^{-10}\,C^2 J^{-1} m^{-1}\right)}$$

$$\times \left(1 - \frac{1}{78.3} - \frac{298\,K}{78.3^2}\,0.356\,K^{-1}\right)\Bigg\} = -375.5\,kJ\,mol^{-1}$$

Correspondingly, the heat of solvation for Ca^{+2} is - **1389.7 kJ mol⁻¹**. The heat of solvation for F^- and Cl^- can be obtained from a similar equation, Eq. (2.48) in the textbook, giving, - **326.1** and - **226.0 kJ mol⁻¹**, respectively.

2.19 Calculate the differences of the heat of hydrations when using the ion-quadrupole model or the ion-dipole model for the ions K^+, Ca^{+2}, F^-, and Cl^-. (Cf. Exercise 2.16 in the textbook) (Kim)

Data:

$$r_{K^+} = 133\,pm \qquad r_{Cl^-} = 181\,pm$$

$$r_{Ca^{+2}} = 99\,pm \qquad r_w = 138\,pm \qquad p_w = 3.9 \times 10^{-10}\,Dm$$

$$r_{F^-} = 136\,pm \qquad \mu_w = 1.87\,D$$

Answer:

The heat of interaction for negative ions and water expressed in the *mksa* system is (cf. Eq. 2.47 in the textbook):

$$\Delta H_{I-S} = 20 - \frac{4N_A z_i e_o \mu_W}{(r_i + r_W)^2\,4\pi\varepsilon_0} + \frac{4N_A z_i e_o p_W}{2(r_i + r_W)^3\,4\pi\varepsilon_0}$$

$$-\frac{N_A (z_i e_o)^2}{2(r_i + 2r_W)\,4\pi\varepsilon_0} \times \left[1 - \frac{1}{\varepsilon} - \frac{T}{\varepsilon^2}\left(\frac{\partial\varepsilon}{\partial T}\right)_P\right] \qquad (2.58)$$

The difference between the quadrupole and the dipole model for water is given by the term $\dfrac{4N_A z_i e_o PW}{2(r_i + r_W)^3 \, 4\pi\varepsilon_0}$. Thus, for K^+,

$$\frac{4N_A z_i e_o PW}{2(r_i + r_W)^3 \, 4\pi\varepsilon_0} = \frac{4 \times 6.023 \times 10^{23}\ mol^{-1}\ (+1)\left(1.602 \times 10^{-19}\ C\right)}{2\left(133 \times 10^{-12}\ m + 138 \times 10^{-12}\ m\right)^3}$$

$$\times \frac{3.9 \times 10^{-10}\ Dm}{\left(1.112 \times 10^{-10}\ C^2 J^{-1} m^{-1}\right)} \times \frac{3.336 \times 10^{-30}\ Cm}{1D} \quad (2.59)$$

$$= 113.4 \text{kJ mol}^{-1}$$

Similarly, the difference of heat of hydration between the dipole and the quadrupole model for Ca^{2+} is **339.1 kJ mol⁻¹**. For anions, the difference between the quadrupole and the dipole model for water is (cf. Eq. 2.48 in textbook) $-\dfrac{4N_A z_i e_o PW}{2(r_i + r_W)^3 \, 4\pi\varepsilon_0}$. Thus, for F^-, this term gives **−108.84 kJ mol⁻¹**, and for Cl^- is **-71.16 kJ mol⁻¹**.

2.20 Calculate the concentration of a NaCl solution at which the so-called "Gurney co-sphere" is reached, that is, when the separation between ions is equal to the radii of the primary solvation sheaths. Consider that for a 1:1 electrolyte, the average separation between ions is given by

$$d = 9.40\, c^{-\frac{1}{3}} \quad (2.60)$$

where d is given in Å and c in mol dm⁻³. (Cf. Exercise 2.7 in the textbook) (Xu)

Answer:

When the average separation between the ions is equal to the radii of primary solvation sheath, then the incompressible spheres are in contact and the Gurney co-sphere is reached. The radii of the primary solvation sheath for the chloride and sodium ions are:

$$r_{h,Na^+} = r_{Na^+} + 2r_w = 0.98 \times 10^{-8} \text{ cm} + 2\left(1.38 \times 10^{-8} \text{ cm}\right)$$

$$= 3.74 \times 10^{-8} \text{ cm}$$

(2.61)

$$r_{h,Cl^-} = r_{Cl^-} + 2r_w = 1.81 \times 10^{-8} \text{ cm} + 2\left(1.38 \times 10^{-8} \text{ cm}\right)$$

$$= 4.57 \times 10^{-8} \text{ cm}$$

(2.62)

The mean separation distance of these ions is $(3.74 + 4.57) \times 10^{-8}$ cm = 8.31 $\times 10^{-8}$ cm. Applying Eq. (2.60) to find the concentration,

$$c = \left(\frac{9.40}{d}\right)^3 = \left(\frac{9.40}{8.31}\right)^3 = 1.45 \text{ M}$$

(2.63)

2.21 Suppose the results from Exercise 2.20 are true. Calculate the solvation number of NaCl. Comment on the reliability of the results. (Cf. Exercise 2.8 in the textbook) (Xu)

Answer:

In Exercise 2.20, all the water molecules in the 1.45 M NaCl solution (i.e., at the most 55.56 mol dm^{-3}) are assumed to be in the primary sphere of the ions. Therefore, the solvation number is:

$$n_s = \frac{55.56 \text{ mol water}}{(2 \times 1.45) \text{ mol ions}} = 19 \text{ !!!}$$

(2.64)

This is a very high value, and indicates that at high concentrations, the average separation relation of Exercise 2.20 oversimplifies the situation. Other effects such as electrostriction should be taken into account in the calculation of n_s.

2.22 A Raman spectrum shows that in a 4.0 M NaCl solution about 40% of the water molecules are in the primary sheath. (a) Estimate the average solvation number, n_s. (b) If the SB region consists of only one layer of water molecules, is there any bulk water left in this solution? (Cf. Exercise 2.9 in the textbook) (Xu)

Data:

$r_{Cl^-} = 181$ nm $r_w = 138$ nm. $c_{NaCl} = 4.0$ mol dm^{-3}

$r_{Na^+} = 95$ nm $n_s \% = 40 \%$

Answer:

(a) The total number of moles of water in the primary sheath in 1 liter of the 4.0 M solution is:

$$solvation\ moles = (n_w)(n_s \%) = (55.56\ \text{mol})(0.40) = 25.00\ \text{mol} \qquad (2.65)$$

Since there are 4.0 moles of electrolyte in 1 liter of this solution, thus, the total solvation number is:

$$n_s = \frac{25.00\ \text{mol water}}{4\ \text{mol electrolyte}} = 6.0 \qquad (2.66)$$

(b) Considering an average ionic radius of $(r_{Na^+} + r_{Cl^-})/2 = (95\ \text{nm} + 181\ \text{nm})/2 = 138\ \text{nm}$, it holds that (cf. Eq. 2.85 in textbook):

$$n_{SB} = \frac{4\pi(r_i + 2r_w)^2}{\pi r_w^2} = \frac{4(138\ \text{nm} + 2(138\ \text{nm}))^2}{138^2\ \text{nm}^2} \qquad (2.67)$$

$$= 36\ \text{water molecules per ion}$$

This means that in 1 liter of solution there are 8 mol of ions present, and thus, the total number of moles of water in the SB regions of the ions is:

$$(8\ \text{mol of ions})\left(36\frac{\text{mol of SB water}}{\text{mol of ion}}\right) = 288\ \text{mol of SB water} \qquad (2.68)$$

Now, 288 mol >> 55.56 mol, which is the number of water moles in 1 liter of water if there were no electrolyte present. Therefore, there is no free bulk water at this concentration. The solution at this point is so concentrated that ions in it have to share water molecules in the SB region.

2.23 Using the equation provided by the Born theory, calculate the free energy of ion-solvent interaction for K^+, Ca^{+2}, F^- and Cl^- in water. The ionic

radii are 133, 99, 136 and 181 pm, respectively, and the dielectric constant of water is 78.3 at 298 K. (Cf. Exercise 2.13 in the textbook) (Kim).

Data:

$$r_{K^+} = 133\,pm \qquad r_{F^-} = 136\,pm$$
$$r_{Ca^{+2}} = 99\,pm \qquad r_{Cl^-} = 181\,pm \qquad \varepsilon_w\,(298\,K) = 78.3$$

Answer:

The energy change due to the solvent interaction with the ion expressed in the *mksa* system is (cf. Eq. A2.1.6 in the textbook):

$$\Delta G_{I-S} = -\frac{N_A\,(z_i e_o)^2}{2r_i\,4\pi\varepsilon_0}\left(1-\frac{1}{\varepsilon}\right) \tag{2.69}$$

For K^+,

$$\Delta G_{K^+-S} = -\frac{6.023\times10^{23}\,mol^{-1}\,(+1)^2\left(1.602\times10^{-19}\,C\right)^2}{2\left(133\times10^{-12}\,m\right)\left(1.112\times10^{-10}\,C^2 J^{-1} m^{-1}\right)}\left(1-\frac{1}{78.3}\right)$$

$$= -515.9\;kJ\;mol^{-1} \tag{2.70}$$

Correspondingly, the ion-solvent interaction free energy for Ca^{+2}, F^- and Cl^- in water are -2772.3, - 504.5, and 379.1 kJ mol^{-1}, respectively.

2.24 (a) Lithium and chloride ions have ionic radius of 59 and 181 pm, respectively. Calculate the work of charging Li^+ and Cl^- in vacuum. **(b)** Calculate now the work of charging Li^+ in water with a relative dielectric constant of 80 at 293 K. (Cf. Exercise 2.28 in the textbook) (Contractor)

Answer:

(a) The work of charging a Li^+ ion in vacuum is (cf. Eq. A2.1.4 in the textbook)

$$W = \frac{N_A (z_i e_o)^2}{8\pi\varepsilon_0 r_i} = \frac{\left(6.022 \times 10^{23}\ \text{mol}^{-1}\right)(+1)^2 \left(1.602 \times 10^{-19}\ \text{C}\right)^2}{2\left(1.112 \times 10^{-10}\ \text{C}^2\text{J}^{-1}\text{m}^{-1}\right)\left(59 \times 10^{-12}\ \text{m}\right)} \quad (2.71)$$

$$= 1178\ \text{kJ mol}^{-1}$$

and the work of charging a Cl⁻ ion in vacuum is:

$$W = \frac{N_A (z_i e_o)^2}{8\pi\varepsilon_0 r_i} = \frac{\left(6.022 \times 10^{23}\ \text{mol}^{-1}\right)(-1)^2 \left(1.602 \times 10^{-19}\ \text{C}\right)^2}{2\left(1.112 \times 10^{-10}\ \text{C}^2\text{J}^{-1}\text{m}^{-1}\right)\left(181 \times 10^{-12}\ \text{m}\right)} \quad (2.72)$$

$$= 384\ \text{kJ mol}^{-1}$$

The work of charging is less in the case of the chloride ion because its radius is larger. Note also that the sign is the same in both cases.

(b) The work of charging a Li⁺ ion in water is (cf. Eq. A2.1.5 in the textbook)

$$W = \frac{N_A (z_i e_o)^2}{8\pi\varepsilon_0 \varepsilon r_i} = \frac{\left(6.022 \times 10^{23}\ \text{mol}^{-1}\right)(+1)^2 \left(1.602 \times 10^{-19}\ \text{C}\right)^2}{2\left(1.112 \times 10^{-10}\ \text{C}^2\text{J}^{-1}\text{m}^{-1}\right)(80)\left(59 \times 10^{-12}\ \text{m}\right)} \quad (2.73)$$

$$= 14.7\ \text{kJ mol}^{-1}$$

A polar medium like water reduces the electrostatic force between two charges by a factor of ε in comparison to the vacuum situation. Hence, *less* work needs to be done to charge the Li⁺ ion in water.

Review of Sections 2.15.12 to 2.15.15 of the Textbook

Describe the entropy changes accompanying hydration according to Bockris and Saluja's model. Write equations for the entropy of (a) an ion in the gas phase, (b) the Born charging, (c) translation of an ion in a liquid, (d) SC water, (e) NSC water, and (f) SB water. Explain which one of the three models, namely, Bernal-Fowler, Frank-Wen or Bockris-Reddy, for hydration entropy best describes experimental facts. What conclusions related to the structure of hydration water can be withdrawn from this analysis? How does the compensation effect of entropy and enthalpy affect the free energy of solvation?

2.25 Using the Born equation as representing a part of the free energy of hydration of ions, derive an expression for the entropy of Born hydration. What would be the entropy of Born hydration of the iodide ion, which radius is 216 pm? Consider that the variation of the dielectric constant with temperature is equal to 0.4 K^{-1}, and that the value of the dielectric constant is 80. (Cf. Exercise 2.12 in the textbook) (Bockris-GamboaAldeco)

Data:

$$r_{I^+} = 216 \text{ pm} \qquad\qquad \varepsilon = 80 \qquad\qquad \partial\varepsilon/\partial T = 0.4$$

Answer:

The free energy according to the Born equation expressed in the *mksa* system is (cf. Eq. A2.1.6 in the textbook):

$$\Delta G_{Born} = -\frac{N_A \left(z_i e_0\right)^2}{\left(4\pi\varepsilon_0\right)2r_i}\left(1-\frac{1}{\varepsilon}\right) \qquad\qquad (2.74)$$

A well known thermodynamic relationship establishes that,

$$\Delta S = -\left(\frac{\partial\Delta G}{\partial T}\right)_P \qquad\qquad (2.75)$$

Therefore,

$$\Delta S_{i-S} = -\left(\frac{\partial\Delta G_{I-S}}{\partial T}\right)_P = -\frac{N_A\left(z_i e_o\right)^2}{2r_i\, 4\pi\varepsilon_0}\frac{1}{\varepsilon^2}\left(\frac{\partial\varepsilon}{\partial T}\right)_P \qquad\qquad (2.76)$$

and the entropy change due to the interaction of I$^-$ with water is

$$\Delta S_{I^- -w} = \frac{6.023\times10^{23}\,\text{mol}^{-1}\,(-1)^2\left(1.602\times10^{-19}\,\text{C}\right)^2}{2\left(216\times10^{-12}\,\text{m}\right)\left(1.112\times10^{-10}\,\text{C}^2\text{J}^{-1}\text{m}^{-1}\right)}\left(\frac{1}{80}\right)^2\,0.4\,\text{K}^{-1}$$

$$= 20.1 \text{ J mol}^{-1} \text{K}^{-1} \tag{2.77}$$

2.26 Calculate the entropy change due to ion-solvent interaction for the ions K^+, Ca^{+2}, F^- and Cl^- in water. Consider the relation $\partial \varepsilon / \partial T = 0.356 \text{ K}^{-1}$ for water. (Cf. Exercise 2.14 in the textbook) (Kim).

Data:

$r_{K^+} = 133 \text{ pm}$	$r_{F^-} = 136 \text{ pm}$	$\varepsilon_w (298 \text{ K}) = 78.3$
$r_{Ca^{+2}} = 99 \text{ pm}$	$r_{Cl^-} = 181 \text{ pm}$	$\partial \varepsilon / \partial T = 0.356 \text{ K}^{-1}$

Answer:

The entropy is calculated from the thermodynamic relationship,

$$\Delta S = -\left(\frac{\partial \Delta G}{\partial T} \right)_P \tag{2.78}$$

Applying this equation to the equation giving the energy change due to solvent interaction with the ion (cf. Eq. A.2.16 in the textbook):

$$\Delta S_{i-S} = -\left(\frac{\partial \Delta G_{I-S}}{\partial T} \right)_P = -\frac{N_A (z_i e_o)^2}{2 r_i \, 4 \pi \varepsilon_0} \frac{1}{\varepsilon^2} \left(\frac{\partial \varepsilon}{\partial T} \right)_P \tag{2.79}$$

The entropy change due to the interaction of K^+ with water is, then,

$$\Delta S_{K^+ -w} = \frac{6.023 \times 10^{23} \text{ mol}^{-1} (+1)^2 \left(1.602 \times 10^{-19} \text{ C} \right)^2}{2 \left(133 \times 10^{-12} \text{ m} \right) \left(1.112 \times 10^{-10} \text{ C}^2 J^{-1} m^{-1} \right)}$$
$$\times \left(\frac{1}{78.3} \right)^2 0.356 \, K^{-1} = 30.3 \text{ J mol}^{-1} \text{K}^{-1} \tag{2.80}$$

Correspondingly, the entropy changes for Ca^{+2}, F^- and Cl^- in water are **162.9, 29.6 and 22.3 J mol^{-1}K^{-1}**, respectively.

2.27 Self diffusion coefficients of certain ions are given in the textbook (e.g., $D_{Li+} = 1.0 \times 10^{-5}$ cm^2 s^{-1} and $D_{I-} = 1.47 \times 10^{-5}$ cm^2 s^{-1}). The diffusion coefficient is related to the rate constant of diffusion by the equation $l^2 = 2Dk^{-1}$. (a) What kind of value for the jumping distance of an ion during diffusion, l, would be reasonable? (b) Calculate the times (τ) the Li$^+$ and the I$^-$ ions reside in one place. Consider $\tau = k^{-1}$. Comment on your results. (Cf. Exercise 2.35 in the textbook) (Bockris-GamboaAldeco)

Data:

From Table 2.26 in the textbook:

$D_{Li+} = 1.0 \times 10^{-5}$ cm^2 s$^-1$ $D_{I-} = 1.47 \times 10^{-5}$ cm^2 s^{-1}

Answer:

(a) The parameter l represents the main distance an ion can jump from one site to another site during the diffusion process. A reasonable value for l would be of the order of the radius or diameter of the considered ion. Thus, in general, the value of l can be taken to be of the order of 100 nm.

(b) From the given equation,

$$l^2 = 2Dk^{-1} = 2D\tau \tag{2.81}$$

where D is the diffusion coefficient, k is the rate constant for diffusion and τ is the mean jump time to cover the mean jump distance (seconds per jump). Since the radius of Li$^+$ is 59 pm, then $l \approx 60$ pm and τ is,

$$\tau = \frac{l^2}{2D} = \frac{\left(60 \times 10^{-10} \text{ cm}\right)^2}{2\left(1.0 \times 10^{-5} \text{ cm s}^{-1}\right)} = 1.8 \times 10^{-12} \text{ s} = 5 \text{ ps} \tag{2.82}$$

In the same way, the radius of the I$^-$ ion is 216 pm. Thus, $l \approx 200$ pm and

$$\tau = \frac{l^2}{2D} = \frac{\left(200 \times 10^{-10} \text{ cm}\right)^2}{2\left(1.47 \times 10^{-5} \text{ cm s}^{-1}\right)} = 13.6 \times 10^{-12} \text{ s} = 13.6 \text{ ps} \tag{2.83}$$

The Li^+ ion, being such a small ion, can move much faster than the much bigger iodide ion, spending less time (small τ) in jumping the mean jump distance.

Review of Sections 2.16 to 2.20 of the Textbook

Define *primary-* and *secondary-hydration numbers*. How are these two quantities distinguished from each other? Name several methods used to determine solvation numbers. What influences do hydration waters have on the different types of molecular orbitals? Draw a schematic of hydration heats of transition-metal ions vs. atomic numbers. What characteristics are present in these plots? Explain. Under what circumstances do second hydration waters bond with greater strength than the first hydration waters? Which are the three computational approaches used to study ionic solvation? Compare their effectiveness to determine solvation numbers. Mention the works of Dang, Malenkov, Heinzinger-Palinkas, and Guardia-Padro. What are the phenomena of *salting-in* and *salting-out*? How can these phenomena be explained in terms of the orientational polarizabilities of the nonelectrolyte and water? Write equations for the primary and secondary solvation numbers. What is Setchenow's constant? Mention the importance of salting-in and salting-out phenomena in industry. What parameters are responsible for the occurrence of *anomalous salting-in*? How is the degree of salting-in affected by the size of the ion? What are *dispersion forces*? How does the presence of these forces affect the solubility of nonelectrolytes in ionic solutions? What is understood by the *hydrophobic aspect of solvation*? Explain how solute-solute interactions lead to hydrophobic effects.

2.28 Calculate the change of solubility of 2,4-dinitrophenol in water due to the primary solvation when 0.1 M NaCl is added to the solution and the solvation number of NaCl is assumed to be 5. (Cf. Exercise 2.25 in the textbook) (Kim)

Answer:

From the equation of change of solubility for non-electrolytes (cf. Eq. 2.137 in textbook):

$$\frac{S - S_0}{S_0} = -\frac{cn_s}{55.55} = -\frac{\left(0.1\,\text{mol}\,\text{dm}^{-3}\right)(5)}{55.55\,\text{mol}\,\text{dm}^{-3}} = -0.009 \qquad (2.84)$$

This equation indicates that the solubility of 2,4-dinitrophenol in water decreases by 0.009 x 100 = **0.9%** when NaCl is added to the solution.

Review of Sections 2.21 to 2.27 of the Textbook .

What is *dielectric breakdown*? When does this phenomenon occur? What is the importance of this phenomenon? How does Szklarczyk's theory explain dielectric breakdown? What is *electrostriction*? Where does the electrostriction take place in a solution? Does electrostriction affect ions and molecules as well as water? How do *polyions* affect the solvent structure? How is hydration in polyions measured? Describe the *reverse micelles* technique to study protein dynamics. What is the relation between the structure of biological materials (e.g., coil-like film and helix) and its content of water? Does water in biological systems have a different structure from water *in vitro*? What do spectroscopic studies say about water in biological systems?

2.29 The definition of compressibility is $\beta = -\dfrac{1}{V}\left(\dfrac{\partial V}{\partial P}\right)_T$. **On the approximated assumption that β is constant with pressure, find V as a function of P. Why is your equation applicable only over a limited range of pressure? (Cf. Exercise 2.22 in the textbook) (Bockris)**

Answer:

From the given equation,

$$\beta = -\frac{1}{V}\left(\frac{\partial V}{\partial P}\right)_T \tag{2.85}$$

At constant temperature, Eq. (2.85) can be written as,

$$\frac{dV}{V} = -\beta dP \tag{2.86}$$

Integrating this equation under the assumption that β is independent of pressure,

$$\int \frac{dV}{V} = -\beta \int dP \quad \text{or} \quad lnV = -\beta P + A_1 \quad \text{or} \quad V = A_2\, e^{-\beta P} \tag{2.87}$$

where A_1 and A_2 are constants. This equation is applicable only over a small range of pressure because β is known to be *dependent* on pressure (cf. Section 2.22.3 in the textbook) up to pressures of the order of 10^9 Pa.

2.30 Calculate the pressure exerted by an Na$^+$ ion on a water molecule in the first hydration shell. Consider the cross section of the ion as the area where the force is applied. (GamboaAldeco)

Data:

r_{Na+} = 95 pm r_w = 138 pm

Answer:

Pressure is defined as force/unit area, or

$$P = \frac{F}{A} \tag{2.88}$$

The force is written in the *mksa* system as (cf. Eq. 2.162 in the textbook)

$$F = \frac{2e_0\mu\cos\theta}{4\pi\varepsilon_0\varepsilon\,r^3} \tag{2.89}$$

Since the angle between the centers of the ion and the water molecule is zero, then $\cos\theta = 1$. Considering the dielectric constant in this region to be equal to 6, and the distance between the ion and the molecule equal to the sum of their radii, i.e., $95 + 138 = 233$ pm, then,

$$F = \frac{2e_0\mu\cos\theta}{4\pi\varepsilon_0\varepsilon r^3} = \frac{2\left(1.602\times10^{-19}\,\text{C}\right)(1.8\,\text{D})}{\left(1.112\times10^{-10}\,\text{C}^2\text{J}^{-1}\text{m}^{-1}\right)(6)\left(233\times10^{-12}\,\text{m}\right)^3} \tag{2.90}$$

$$\times\frac{3.336\times10^{-30}\,\text{C m}}{1\,\text{D}}\times\frac{1\,\text{N m}}{1\,\text{J}} = 2.28\times10^{-10}\,\text{N}$$

The area where the force is applied is $\pi r^2 = \pi(95\times10^{-12}\,\text{m})^2 = 2.83\times10^{-20}\,\text{m}^2$. Therefore, the pressure is

$$P = \frac{2.28 \times 10^{-10} \, N}{2.83 \times 10^{-20} \, m^2} \times \frac{1 \, Pa}{1 \, N \, m^{-2}} = 8.04 \times 10^9 \, Pa \qquad (2.91)$$

2.31 Calculate the change of solubility of ethyl ether in water due to the secondary solvation shell, when 1 M NaCl solution is added to the solution. The dielectric constants of ethyl ether and 1 M NaCl solution are 4.33 and 70, respectively, the density of ethyl ether is 0.7138 g cm^{-3}, and the polarizability of water is 1.46 x x10^{-24} cm^3. The polarizability of non-electrolytes can be determined from the following equation:

$$\alpha_{NE} = \frac{(\varepsilon_{NE} - 1)}{4 \pi N} \qquad (2.92)$$

where N is the number of molecules of the non-electrolyte (NE) in one cm^3. (Cf. Exercise 2.26 in the textbook) (Kim)

Data:

$\varepsilon_{ethyl\text{-}ether} = 4.33$	$\varepsilon_{NaCl} = 70$	$\alpha_w = 1.46 \times 10^{-24}$ cm^3
$\rho_{ethyl\text{-}ether} = 0.7138$ g cm^{-3}	$c_{NaCl} = 1$ mol dm^{-3}	$r_{Na+} = 95$ pm
$r_{Cl-} = 181$ pm	$r_w = 138$ pm	

Answer:

The change of solubility of a nonelectrolyte in water due to the secondary solvation shell can be calculated from (cf. Eq. 2.151 in textbook):

$$\frac{S - S_0}{S_0} = \frac{N_A c \, 4\pi (z_i e_0)^2 (\alpha_w - \alpha_{NE})}{1000 \qquad \varepsilon^2 \, kTr_h} \qquad (2.93)$$

To determine the polarizability of the ethyl ether it is needed to determine first N, the number of ethyl ether molecules in 1 cm^3,

$$N = \frac{\rho_{ethyl} N_A}{(MW)_{ethyl}} = \frac{\left(0.7138 \, \text{g cm}^{-3}\right)\left(6.022 \times 10^{23} \, \text{mol}^{-1}\right)}{74.13 \, \text{g mol}^{-1}} \qquad (2.94)$$

$$= 5.8 \times 10^{21} \, \text{cm}^{-3}$$

From Eq. (2.92)

$$\alpha_{ethyl} = \frac{4.3 - 1}{4\pi\left(5.8 \times 10^{21} \ cm^{-3}\right)} = 4.53 \times 10^{-23} \ cm^3 \qquad (2.95)$$

Finally, r_h is

$$r_h = r_i + 2r_w \qquad (2.96)$$

Considering that $r_i \cong (r_{Na+} + r_{Cl-})/2 = 185$ pm, then,

$$r_h = 138 \ pm + 2(138 \ pm) = 414 \ pm = 4.14 \times 10^{-8} \ cm \qquad (2.97)$$

Substituting Eqs. (2.95) and (2.97) into Eq. (2.93)

$$\frac{S - S_0}{S_0} = -\frac{\left(6.022 \times 10^{23} \ mol^{-1}\right)\left(1 \ mol \ dm^{-3}\right)}{1000 \ cm^3 \ dm^{-3}}$$

$$\times \frac{(1)^2 \left(1.602 \times 10^{-19} \ C\right)^2 (1.46 - 45.3) \times 10^{-24} \ cm^3}{\left(8.854 \times 10^{-12} \ C^2 J^{-1} m^{-1}\right)(70)^2 \left(1.381 \times 10^{-16} \ erg \ K^{-1}\right)} \qquad (2.98)$$

$$\times \frac{1}{(298 \ K)\left(4.14 \times 10^{-10} \ m\right)} \times \frac{10^7 \ erg}{1 J} = 0.009$$

2.32 Determine the dielectric constant of a 50% amine solution. Consider the dielectric constant of the pure amine to be 4.3, and that of pure water 78. What is the dielectric constant of a solution 30% of this amine? (GamboaAldeco)

Answer:

The relationship between dielectric constant and concentration is a linear one (cf. Eq. 2.170 in the textbook),

$$\varepsilon = \varepsilon_0 \left(1 - \delta c\right) \qquad (2.99)$$

At $c = 0$, $\varepsilon = \varepsilon_0 = 78$. And at $c = 1$, $4.3 = \varepsilon_0 (1 - \delta)$, or $\delta = 0.95$. Thus,

$$\varepsilon = 78(1 - 0.95c) \tag{2.100}$$

At 50% concentration of amine, $\varepsilon = 78[1 - 0.95(0.50)] = 41$, and at 30%, $\varepsilon = 78[1 - 0.95(0.30)] = 56$.

PROBLEMS

2.33 For a given water molecule, what is the *maximum* number of hydrogen bonds that can be form with other neighboring water molecules? Are these hydrogen bonds identical in bonding nature? Explain. If there were a difference in bonding nature for these hydrogen bonds, how would you differentiate them experimentally? The word *maximum* is emphasized because in liquid state the actual number of hydrogen bonds per molecule is less. Even in the crystal ice-II there are dangling hydrogen atoms that do not participate in hydrogen bonding. (Cf. Problem 2.17 in the textbook) (Xu)

Answer:

One single water molecule can form at the most four hydrogen bonds with other water molecules. They are of two different types: two can be formed with the hydrogen atoms of the reference water molecule, and the other two with the lone pairs of its oxygen atom. The former two are electron-acceptor and the latter two are electron donor. These two types of hydrogen bonds can be distinguished by the O-H distance: 99 *nm* for the H...H type, and 177 nm for the H...O type (cf. Fig. 2.9 in textbook)

2.34 (a) Calculate the dipole moment associated with water in the hypothesis of fully ionic O-H bondings, i.e., considering a $-2e_0$ charge on the O atom and a $+1e_0$ charge on each H atom. Draw the associated vector. Refer to the drawing in Fig. 2.2 for angles and distances in the water molecule. (b) Knowing the dipole moment of water, write an equation expressing it in terms of the fractional charge on either end of the dipole, f, and the distance between the center of charge and the atomic nucleus, x. (c) Determine f and x in the hypothesis of a tetrahedrical angle (i.e., 120^0) between H-O-H, and draw the corresponding dipole moment vector. (Cf. Problem 2.24 in the textbook) (Mussini)

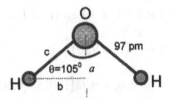

Fig. 2.2. Angles and distances in the water molecule.

Answer:

(a) The dipole moment is given by $\mu = qd$. Under the hypothesis presented here, the dipole moment can be represented as in Fig. 2.3(a), and it is calculated as:

$$\mu = qa = 2e_0 c \cos\frac{\theta}{2} = 2\left(1.602\times10^{-19}\ C\right)\left(97\times10^{-10}\ cm\right)\cos\frac{105}{2} \quad (2.101)$$

$$= 1.89\times10^{-27}\ C\,cm$$

(b) From the dipole moment of water, 1.87 D,

$$\mu_w = 2e_0 fx = 1.87\,D\frac{3.336\times10^{-28}\ C\,cm}{1D} \quad (2.102)$$

Solving for the parameter f,

$$f = \frac{1.94\times10^{-9}}{x} \quad (2.103)$$

with x given in cm. See Fig. 2.3(b).

(c) With the angle between H-O-H equal to 120^0, a new set of variables (a', b' and c') is assigned. With $b = b'$, and $a' = x$, the distance a' is

$$a' = x = b'\tan\left(120/2\right) = 46\ pm \quad (2.104)$$

and the parameter f is,

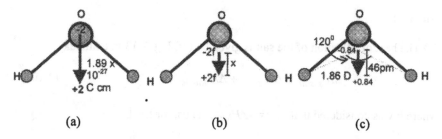

Fig. 2.3. The parameters μ and f in the water molecule according to Problem 2.34.

$$f = \frac{1.94 \times 10^{-9}}{46 \times 10^{-10} \text{ cm}} = 0.42 \qquad (2.105)$$

Finally, the dipole moment becomes,

$$\mu_w = 2 e_0 fx = 2\left(1.602 \times 10^{-19} \text{ C}\right)(0.42)\left(46 \times 10^{-10} \text{ cm}\right)$$

$$= 6.19 \times 10^{-28} \text{ C cm} = 1.86 D \qquad (2.106)$$

See Fig. 2.3(c).

Comment This problem follows the 1933 Bernal and Fowler approach, which is still very appreciable. However, more complicated charge distributions, e.g., quadrupole ones, have been proposed in latter years.

2.35 (a) Using data for the solution enthalpy (ΔH_{soln}) and lattice enthalpy ($\Delta H_{lattice} = -L_{sub}$) in the Table below, calculate the hydration heats for various alkali halides (ΔH_s). Comment on the possible source of major error. (b) Using cation radii data, explain the trend and the sign of ΔH_{soln}. What is the driving force of the solution process when $\Delta H_{soln} > 0$? (Cf. Problem 2.11 in the textbook) (Xu)

	LiCl	LiBr	NaCl	NaBr	KCl	KBr
ΔH_{soln} (kJ mol^{-1})	-37.0	-48.8	+3.89	-0.6	+17.22	+19.9
ΔS_{soln} (kJ mol^{-1} K^{-1})	+48	+43	+110	+100	+135	+140
$\Delta H_{lattice}$ (kJ mol^{-1})	-852	-815	-787	-752	-717	-689

Answer:

(a) The heat of solvation of the salt is given by (cf. Fig. 2.13 in textbook):

$$\Delta H_{s,salt} = \Delta H_{soln} + \Delta H_{lattice} \tag{2.107}$$

where it was considered that $L_{sub} = -\Delta H_{lattice}$. Thus, for LiCl,

$$\Delta H_{s,LiCl} = -37.0 \, kJ \, mol^{-1} - 852 \, kJ \, mol^{-1} = 889 \, kJ \, mol^{-1} \tag{2.108}$$

In the same way for the other salts:

	LiCl	LiBr	NaCl	NaBr	KCl	KBr
$\Delta H_{s,salt}$ (kJ mol^{-1})	-889	-864	-783	-753	-700	-669

Since $\Delta H_{s,salt}$ has dominant contribution from $\Delta H_{lattice}$, the major possible source of error should come from this parameter. Besides, ΔH_{soln} can be accurately measured.

(b) In the solution process: $MX(s) \rightarrow M^{+}(aq) + X^{-}(aq)$, the parameter that indicates whether or not the salt dissolves is ΔG_{soln}. The more negative ΔG_{soln} is, the more stable the ions are in solution. Now, ΔG_{soln} has two contributions, one from the enthalpy of solution, ΔH_{soln}, and another from the entropy of solution, ΔS_{soln}, i.e., $\Delta G_{soln} = \Delta H_{soln} - T\Delta S_{soln}$.

The first term, ΔH_{soln}, is the result of the *stabilization* of the ions due to the ion-solvent interaction ($\Delta H_{s,salt}$), and the *de-stabilization* of the ions due to disruption of the lattice ($\Delta H_{lattice}$). The major contributors to the stabilization factor (Born charging term, ion-dipole/quadrupole terms and induced dipole terms) are proportional to r_i^{-n}, with $n = \{1...4\}$ and r being the radius of the ion. Thus, the smaller the ion, the stabilization of the ions due to solvent-ion interactions becomes greater, and therefore, ΔH_{soln} becomes more negative. A large negative ΔH_{soln} means a large negative ΔG_{soln}.

In the opposite case, i.e., when very large cations are present, e.g. K^{+}, the stabilization due to solvent ion-interactions cannot cover the de-stabilization due to the disruption of the lattice. This results in small negative numbers or even positive values for ΔH_{soln}. In such cases ΔS_{soln} becomes the sole driving force in the dissolution process, because only a large ΔS_{soln} will assure that $\Delta G_{soln} < 0$ and therefore that the salt dissolves.

2.36   **Using the data in the table below, calculate the heats of solution, ΔH_{soln}, and the heats of solvation, ΔH_s, for AgCl, AgBr and AgI and rationalize their low solubility. (Cf. Problem 2.7 in the textbook) (Xu)**

Answer:

The dissolution reaction of a salt in solution is:

	AgCl	AgBr	AgI
K (equilibrium constant)	1.77×10^{-10}	7.7×10^{-13}	1.5×10^{-16}
ΔS_{soln} (kJ mol^{-1} K^{-1})	32.98	51.34	73.59
$\Delta H_{lattice}$ (kJ mol^{-1})	-912	-900	-886

$$MX(s) \rightarrow M^+(aq) + X^-(aq) \qquad (2.109)$$

Its equilibrium constant is

$$K = \left[M^+(aq) \right]\left[X^-(aq) \right] = exp\left(-\frac{\Delta G_{soln}}{RT} \right) \qquad (2.110)$$

and the free energy of solution is given by,

$$\Delta G_{soln} = -RT \ln K \qquad (2.111)$$

The free energy of solution can be split into an enthalpy term and an entropy term:

$$\Delta G_{soln} = \Delta H_{soln} - T\Delta S_{soln} \quad \text{or} \quad \Delta H_{soln} = \Delta G_{soln} + T\Delta S_{soln} \quad (2.112)$$

In addition, the heat of sublimation (L_{sub}) is equal to the negative of the lattice enthalpy ($\Delta H_{lattice}$). Thus, the heat of solvation of the salt is (cf. Fig. 2.13 in the textbook),

$$\Delta H_{s,salt} = \Delta H_{soln} + \Delta H_{lattice} \qquad (2.113)$$

Thus, for AgCl, the free energy of solution from Eq. (2.111) is

$$\Delta G_{soln} = -\left(8.31 \text{ J mol}^{-1} \text{K}^{-1} \right)\left(298 \text{ K} \right)\ln 1.77 \times 10^{-10}$$

$$= 55.61 \text{ kJ mol}^{-1} \tag{2.114}$$

and the enthalpy of solution, from Eq. (2.222) is,

$$\Delta H_{soln} = 55.61 \text{ kJ mol}^{-1} + \left(298 \text{ K} \right)\left(32.98 \times 10^{-3} \text{ kJ mol}^{-1} \text{K}^{-1} \right)$$

$$= 65.44 \text{ kJ mol}^{-1} \tag{2.115}$$

The heat of solvation from Eq. (2.113) is,

$$\Delta H_{s,salt} = 65.44 \text{ kJ mol}^{-1} - 912 \text{ kJ mol}^{-1} = -846.5 \text{ kJ mol}^{-1} \tag{2.116}$$

In the same way, for the other two salts:

	AgCl	AgBr	AgI
ΔG_{soln} (kJ mol^{-1})	55.61	69.10	90.27
ΔH_{soln} (kJ mol^{-1})	65.44	84.40	112.20
$\Delta H_{s,salt}$ (kJ mol^{-1})	-846.5	-815.6	-773.8

Since $\Delta G_{soln} > 0$, then the three salts are insoluble, with AgI being the most insoluble salt. In the three cases $\Delta H_{soln} > 0$, and the entropy change, ΔS_{soln} is not positive enough to make the process happen (see table above). The stable lattice of these salts —marked by the large negative value of $\Delta H_{lattice}$— makes it difficult for the solvent to grab the ions away.

2.37 The densities of aqueous NaCl solutions at 25 ^{0}C are given as a function of NaCl molality in the table below. The density of pure NaCl is 2.165 g cm^{-3}. Calculate the partial molar volumes of : (a) NaCl ($\tilde{v}_2$) , and (b) water ($\tilde{v}_1$),and plot these values as a function of NaCl molality. (c) Compare the limiting cases, i.e., $\tilde{v}_1$ when $m \to 0$, and $\tilde{v}_2$ when $m \to$ saturation, with those of pure water and pure NaCl molar volumes, V_1^0 and

V_2^0, respectively. (d) Calculate $\tilde{V}_2$ for $m = 0.5$ mol kg^{-3} and $m = 2$ mol kg^{-3}. (Cf. Problem 2.23 in the textbook) (Mussini)

m_{NaCl} (mol kg^{-1})	ρ (kg dm^{-3})	m_{NaCl} (mol kg^{-1})	ρ (kg dm^{-3})
0	0.99707	3.09392	1.10849
0.11094	1.00158	3.9873	1.13600
0.23631	1.00663	5.24324	1.17290
0.56874	1.01970	5.4952	1.18030
0.85382	1.03071	5.8023	1.18880
1.47458	1.05353	5.82267	1.18888
2.51393	1.08963		

Answer:

(a) The volume of the solution can be divided into (cf. Eq. 2.6 in the textbook):

$$V = n_1 V_{m,1} + n_2 V_{m,2} \tag{2.117}$$

where $V_{m,2}$ is the *apparent* NaCl molar volume in the solution and $V_{m,1}$ is the molar volume of the solvent. The apparent molar volume is a quantity that can be obtained from the experimental values of density of the solution. Differentiating Eq. (2.117) with respect to n_2, keeping n_1 constant,

$$\left(\frac{\partial V}{\partial n_2} \right)_{n_1} = n_2 \left(\frac{\partial V_{m,2}}{\partial n_2} \right)_{n_1} + V_{m,2} \tag{2.118}$$

Now, the total volume of the solution can also be divided into:

$$V = n_1 \tilde{V}_1 + n_2 \tilde{V}_2 \tag{2.119}$$

where $\tilde{V}_1$ and $\tilde{V}_2$ are the *partial* molar volumes of water and NaCl, respectively. However, the *partial* molar volumes cannot be determined directly from experimental data, although can be related to the corresponding *apparent* molar volumes. The partial molar volume is defined as $(\partial V / \partial n_i)_{n_j}$. Therefore, $\tilde{V}_2 = (\partial V / \partial n_2)_{n_1}$. Then, making use of Eq. (2.118), the partial molar volume of NaCl ($\tilde{V}_2$) can be written in terms of the apparent molar volume of NaCl ($V_{m,2}$) as:

$$\tilde{V}_2 = n_2 \left(\frac{\partial V_{m,2}}{\partial n_2} \right)_{n_1} + V_{m,2} \tag{2.120}$$

The condition of constant n_1 can be observed by considering always an amount of solution corresponding to 1 kg of solvent, i.e., $n_1 = 1000$ g /18.0152 g mol^{-1} = 55.51 mol, an advantageous conditions because in this case $n_2 = m$.

The parameter, $V_{m,2}$ can be obtained from Eq. (2.117) (cf. Eq. 2.6 in the textbook:

$$V_{m,2} = \frac{V - n_1 V_{m,1}}{n_2} = \frac{1}{n_2} \left[V - n_1 \frac{m_1}{\rho_1 n_1} \right] = \frac{1}{n_2} \left[V - \frac{m_1}{\rho_1} \right] \tag{2.121}$$

The total volume of the solution, V, can be obtained from the data of solution density. Thus,

$$V = \frac{m_{soln}}{\rho_{soln}} = \frac{m_w + m_{NaCl}}{\rho_{soln}} = \frac{m_w + n_2 (MW)_{NaCl}}{\rho_{soln}}$$

$$= \frac{1 kg + n_2 \left(0.0585 \, kg \, mol^{-1} \right)}{\rho_{soln}} \tag{2.122}$$

For $c_{NaCl} = 0.11094$ mol kg^{-1} and $\rho = 1.00158$ kg dm^{-3}, Eq. (2.122) becomes,

$$V = \frac{1 kg + \left(0.11094 \, mol \right) \left(0.0585 \, kg \, mol^{-1} \right)}{\left(1.00158 \, kg \, dm^{-3} \right)} = 1.0049 \, dm^{-3} \tag{2.123}$$

and Eq. (2.121) becomes,

$$V_{m,2} = \frac{1}{0.11094 \, mol} \left[1.00490 \, dm^3 - \frac{1 kg}{0.99707 \, kg \, dm^{-3}} \right] \tag{2.124}$$

$$= 0.01770 \, dm^3 \, mol^{-1}$$

The values of V and $V_{m,2}$ at different concentrations are shown in the following table for 1 kg of solvent:

Molality[a]	Density[b]	V[c]	$V_{m,2}$[d]	$dV_{m,2}/dn_2$[e]	$\tilde{V}_2$[f]	$\tilde{V}_1$[g]
0 (water)	0.99707					
0.11094	1.00158	1.0049	0.01770	0.00103	0.0178	0.0181
0.23631	1.00663	1.0071	0.01781	0.00101	0.0181	0.0181
0.56874	1.0197	1.0133	0.01823	0.00096	0.0187	0.0181
0.85382	1.03071	1.0187	0.01842	0.00092	0.0192	0.0181
1.47458	1.05353	1.0311	0.01908	0.00084	0.0202	0.0180
2.51393	1.08963	1.0527	0.01980	0.00070	0.0215	0.0180
3.09392	1.10849	1.0654	0.02019	0.00062	0.0221	0.0180
3.9873	1.1366	1.0850	0.02059	0.00050	0.0226	0.0179
5.24324	1.17290	1.1141	0.02120	0.00032	0.0229	0.0179
5.4952	1.18030	1.1196	0.02123	0.00029	0.0229	0.0179
5.8023	1.18880	1.1267	0.02133	0.00025	0.0228	0.0179
5.82267	1.18888	1.1276	0.02142	0.00024	0.0228	0.0179
NaCl	2.165					

[a] mol_{NaCl}/kg_{solv} [c] dm^3 [e] dm^3/mol^2 [g] dm^3/mol
[b] kg_{solv}/dm^3_{soln} [d] dm^3/mol [f] dm^3/mol

The next step is to find the value of $\left(\partial V_{m,2}/\partial n_2 \right)_{n_1}$. This can be done from the slope of a $V_{m,2}$ vs. n_2 curve. Plotting $V_{m,2}$ against n_2 (or m) gives the graph in Fig. 2.4. The data fit the second-degree polynomial:

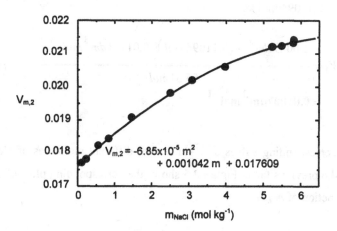

$$V_{m,2} = -6.85 \times 10^{-5} m^2 + 0.001042\, m + 0.017609$$

Figure 2.4. The apparent NaCl molar volume vs. molality

$$V_{m,2} = -0.0000685\,n_2^2 + 0.001042\,n_2 + 0.017609 \text{ in } dm^3\,mol^{-1} \quad (2.125)$$

Differentiating this equation with respect to n_2 (or m) gives:

$$\left(\frac{\partial V_{m,2}}{\partial n_2}\right)_{n_1} = -0.0001370\,n_2 + 0.001042 \text{ in } dm^3\,mol^{-2} \quad (2.126)$$

Now we have the parameters to find $\tilde{V}_2$. Thus, substituting the corresponding values into Eq. (2.120) for $m = 0.11094$ mol kg^{-1} and using the values of $V_{m,2}$ calculated before,

$$\tilde{V}_2 = 0.11094\ mol\left\{[-0.0001370(0.11094) + 0.001042]dm^3\,mol^{-2}\right\} \\ + 0.01770\,dm^3\,mol^{-1} = 0.01781\,dm^3\ mol^{-1} \quad (2.127)$$

(b) Once the values of the partial-molar volume of the solute are known, the partial-molar volume of water, $\tilde{V}_1$, can be calculated from Eq. (2.119), i.e.,

$$\tilde{V}_1 = \frac{V - n_2\tilde{V}_2}{n_1} \quad (2.128)$$

For $m_{NaCl} = 0.11094$ mol kg^{-1},

$$\tilde{V}_1 = \frac{1.0049\,dm^3 - (0.11094\,mol)\left(0.01781\,dm^3\,mol^{-1}\right)}{55.51\,mol} \quad (2.129)$$
$$= 0.01807 dm^3\,mol^{-1}$$

The corresponding values of $\tilde{V}_2$ and $\tilde{V}_1$ at different molalities of NaCl are given in the previous table. Figure 2.5 shows the corresponding plots of $\tilde{V}_2$ and $\tilde{V}_1$ as a function of m_{NaCl}.

(c) Limiting case I: $m \to 0$, $\tilde{V}_1 \to V^0_{water}$

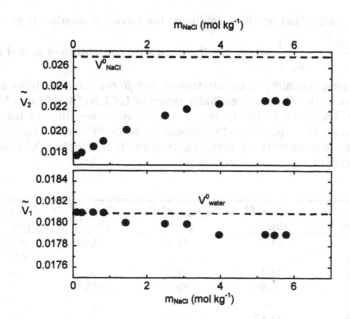

Figure 2.5. Partial-molar volumes of water ($\tilde{V}_1$) and of NaCl ($\tilde{V}_2$) vs. the molarity of the solution.

$$V_{water}^0 = \frac{MW_{water}}{\rho_{water}} = \frac{18.016 \ \text{g mol}^{-1}}{997.07 \ \text{g dm}^{-3}} = 0.0181 \ \text{dm}^{-3} \ \text{mol}^{-1} \qquad (2.130)$$

This value is shown in Fig. 2.5. There is a good agreement between $(\tilde{V}_1)_{m \to 0}$ and V_{water}.

Limiting case II: $m \to$ sat, $\tilde{V}_2 \to V_{NaCl}^0$

$$V_{NaCl}^0 = \frac{MW_{NaCl}}{\rho_{NaCl}} = \frac{58.50 \ \text{g mol}^{-1}}{2165 \ \text{g dm}^{-3}} = 0.02702 \ \text{dm}^{-3} \ \text{mol}^{-1} \qquad (2.131)$$

This value is shown in Fig. 2.5. It shows that $(\tilde{V}_2)_{m \to \infty} < V_{NaCl}$.

(d) $(\tilde{V}_2)_{m = 0.5 \, \text{mol kg}^{-1}} = 18.6 \ \text{cm}^3 \ \text{mol}^{-1}$ and $(\tilde{V}_2)_{m = 2.0 \, \text{mol kg}^{-1}} = 21.0 \ \text{cm}^3 \ \text{mol}^{-1}$.

2.38 (a) Justify that in dilute solutions, the solvation number is given by

$$n_s = \frac{55.56}{c}\left(1 - \frac{\beta}{\beta_0}\right),$$ **where c is the electrolyte concentration in mol dm^{-3},**

β **is the compressibility of the electrolyte, and β_0 the compressibility of the pure solvent. (b) The compressibility values of LiCl, NaCl, KCl and MgCl$_2$ solutions are given in the table below. The compressibility of the pure water, under the experimental conditions, is 44.65×10^{-6} bar^{-1}. Calculate the total solvation number of these electrolytes. (Cf. Problem 2.21 in the textbook) (Xu)**

c (mol dm^{-3})	$\beta_{LiCl} \times 10^6$ (bar^{-1})	$\beta_{NaCl} \times 10^6$ (bar^{-1})	$\beta_{KCl} \times 10^6$ (bar^{-1})	$\beta_{MgCl2} \times 10^6$ (bar^{-1})
0.05	44.49	44.45	44.46	44.21
0.09	-	-	-	43.88
0.10	44.30	44.20	-	-
1.00	40.60	40.04	39.84	-
1.05	-	-	-	35.94
2.00	37.15	36.64	-	-
4.10	-	-	29.88	-
5.00	29.93	26.75	-	-

Answer:

(a) The volume of an aqueous solution can be divided into two parts according to its compressibility: a compressible volume and an incompressible volume.

The incompressible volume, v, consists of the volume of water molecules that are in the primary shell. If the number of water molecules around a single ion is n_s, and c is the concentration of ions in mol liter^{-1}, then, the total number of water molecules in the primary shell is cn_s. Thus, the incompressible volume of water is $v = V_0 cn_s$, where V_0 is the molar volume of water. From the definition of compressibility (cf. Eq. 2.9 in the textbook):

$$\beta = -\frac{1}{V}\left[\frac{\partial(V - v)}{\partial P}\right]_T = -\frac{1}{V}\left(\frac{\partial V}{\partial P}\right)_T \qquad (2.132)$$

where V is the total volume of solution given by $V \approx 55.56 V_0$ (for diluted solutions), and $(V-v)$ is the volume of the compressible water. However, this compressible water should have the compressibility value of pure water, i.e., β_0:

$$\beta_0 = -\frac{1}{V-v}\left[\frac{\partial(V-v)}{\partial P}\right]_T = -\frac{1}{V-v}\left(\frac{\partial V}{\partial P}\right)_T \qquad (2.133)$$

Therefore,

$$\frac{\beta}{\beta_0} = \frac{V-v}{V} = 1 - \frac{v}{V} = 1 - \frac{V_0 cn_s}{55.56V_0} = 1 - \frac{cn_s}{55.56} \qquad (2.134)$$

or

$$n_s = \left(1 - \frac{\beta}{\beta_0}\right)\frac{55.56}{c} \qquad (2.135)$$

(b) Using the above relationship, for 5.00 M LiCl,

$$n_s = \left(1 - \frac{29.93 \times 10^{-6}\, bar^{-1}}{44.65 \times 10^{-6}\, bar^{-1}}\right)\frac{55.5\,mol\,dm^{-3}}{5.00\,mol\,dm^{-3}} = 3.7 \qquad (2.136)$$

The values of total solvation number of the other electrolytes are listed next:

c(mol dm^{-3})	LiCl	NaCl	KCl	MgCl$_2$
0.05	4.0	5.0	4.7	11.0
0.09	-	-	-	11.0
0.10	4.8	5.6	-	-
1.00	5.0	5.7	6.0	-
1.05	-	-	-	10.3
2.00	4.7	5.0	-	-
4.10	-	-	4.5	-
5.00	3.7	4.5	-	-

2.39 (a) Calculate the transport numbers, t_+, of Li$^+$, Na$^+$, K$^+$, and Mg^{2+} in dilute chloride solutions. Use the mobility data, μ_+ and μ_- in the following equations and table:

$$t_+ = \frac{\mu_+}{\mu_+ + \mu_-} \qquad (2.137)$$

and

$$t_+ + t_- = 1 \tag{2.138}$$

Ion	Li^+	Na^+	K^+	Mg^{+2}	Cl^-
$\mu\,(10^{-8}\ m^2\ s^{-1}V^{-1})$	4.01	5.19	7.62	5.42	7.91

(b) Estimate the term A in the following equation that relates the solvation numbers of the ions in the chloride salts based on their transport numbers (cf. Eq. 2.17 in textbook):

$$\left(\frac{t_+}{|z_+|}\right)n_+ - \left(\frac{t_-}{|z_-|}\right)n_- = \frac{\Delta E}{2.79\,a_0} + A \tag{2.139}$$

Use the values of $(V_s)_+$ and $(V_s)_-$ given in Table 2.6 in the textbook. (Xu)

Answer:

(a) Using the Eq. (2.137) for transport number for the first solution, LiCl,

$$t_+ = \frac{\mu_+}{\mu_+ + \mu_-} = \frac{4.01}{4.01 + 7.91} = 0.33 \tag{2.140}$$

In the same way, $t_{Na^+} = 0.39$; $t_{K^+} = 0.49$; $t_{Mg^{2+}} = 0.41$

(b) Comparing Eq. (2.16) in the textbook with Eq. (2.139)

$$A = \frac{\rho_1}{M_1}\left(\frac{t_+}{|z_+|}(V_s)_+ - \frac{t_-}{|z_-|}(V_s)_-\right) - \frac{1}{M_1}\left(\frac{t_+}{|z_+|}M_+ - \frac{t_-}{|z_-|}M_-\right) \tag{2.141}$$

Substituting the corresponding values for LiCl,

$$A_{LiCl} = \frac{1}{18}\left(\frac{0.33}{1}(140.7) - \frac{1-0.33}{1}(92.4)\right) \tag{2.142}$$

$$-\frac{1}{18}\left(\frac{0.33}{1}(6.9)-\frac{1-0.33}{1}(35.5)\right)=0.34$$

In the same way for the other electrolytes: A_{NaCl} = 0.082; A_{KCl} = -0.18; A_{MgCl2} = 0.12.

2.40 Suppose that the transport numbers obtained in Problem 2.39 are independent of concentration. Calculate the absolute ion solvation number, n_+ and n_-, for (a) LiCl and (b) MgCl$_2$. Use the value of $\Delta E/2.79\,a_0$ from Table 2.6 in the textbook, the value of A obtained from Problem 2.39, and the total solvation number n_s, given by $n_s = |z_+|n_+ + |z_-|n_-$, obtained in Problem 2.38. (Xu)

Answer:

(a) From Problem 2.38, n_s for a solution of 0.05M LiCl is $n_{s,LiCl}$ = 4.0. Thus,

$$n_s = |z_+|n_+ + |z_-|n_- = n_+ + n_- = 4.0 \tag{2.143}$$

Substituting this equation in the equation that relates the solvation numbers of the ions in salts based on their transport numbers (cf. Eq. 2.17 in the textbook):

$$\left(\frac{t_+}{|z_+|}\right)n_+ - \left(\frac{t_-}{|z_-|}\right)n_- = \frac{\Delta E}{2.79\,a_0} + A \tag{2.144}$$

Rearranging terms in Eq. (2.144),

$$\left(\frac{t_+}{|z_+|}\right)(n_s - n_-) - \left(\frac{1-t_+}{|z_-|}\right)n_- = \frac{\Delta E}{2.79\,a_0} + A \tag{2.145}$$

Substituting the parameters A = 0.34 from Problem 2.39 and $\Delta E/2.79a_0$ = 0.1/2.79 from Table 2.6 in the textbook, and solving for n_-,

$$(0.33)(4.0-n_-)-(1-0.33)n_- = \frac{0.1}{2.79}+0.34 \tag{2.146}$$

Solving for n_- gives, $n_- = 0.94$. Also, $n_+ = n_s - n_- = 4.0 - 0.94 = 3.1$

(b) Similarly, for 1 M $MgCl_2$, $n_{s,MgCl2} = 10.3$ from Problem 2.38:

$$n_s = n_+ + 2n_- = 10.3 \tag{2.147}$$

Considering the parameters $A = 0.12$ from Problem 2.39 and $\Delta E/2.79a_0 = 1/2.79$ from Table 2.6 in the textbook, Eq. (2.145) for $MgCl_2$ is written as

$$\frac{0.41}{2}(10.3 - 2n_-) - (1 - 0.41)n_- = \frac{1.0}{2.79} + 0.12 \tag{2.148}$$

Solving for n_- and n_+ gives $n_- = 1.63$ and $n_+ = 10.3 - 2(1.63) = 7.04$

	LiCl		MgCl$_2$	
	Li$^+$	Cl$^-$	Mg^{+2}	Cl$^-$
n_+	3.1	0.94	7.04	1.63

2.41 The following table lists the measured dielectric constants at 25 ^{0}C for 1.0 M LiCl, NaCl, and KCl solutions. Calculate the percentage of water in the primary sheath. Compare your results with those of the compressibility method and comment on their reliability. (Cf. Problem 2.22 in the textbook) (Xu)

	LiCl	NaCl	KCl
ε at 25 ^{0}C	64.9	66.7	68.1

Answer:

From the model of Hasted et al., (cf. Eq. 2.27 in the textbook),

$$\varepsilon = 80\frac{55.56 - c_i n_s}{55.56} + 6\frac{c_i n_s}{55.56} \tag{2.149}$$

where 55.56 stands for the number of moles of water in 1 dm^3, and c_i is given in mol dm^{-3}. Thus, solving for n_s,

$$n_s = \frac{55.56(80-\varepsilon)}{74c_i} \tag{2.150}$$

Therefore, the primary hydration number for LiCl is,

$$n_s(LiCl) = \frac{55.56(80-64.9)}{74(1.0)} = 11 \tag{2.151}$$

This result indicates that about 11/55.56 or **20%** of water molecules are in the primary sheath. In the same way for NaCl and KCl, about 10/55.56 or **18%** and 9/55.56 or **16%** of water molecules respectively are in the primary sheath. Comparing these values with the corresponding ones obtained from the adiabatic compressibility method (cf. Table 2.5 in textbook: $n_s(LiCl) = 4.9$, $n_s(NaCl) = 5.7$, and $n_s(KCl) = 5.7$, the solvation numbers obtained through the dielectric constant method are higher.

The major error source is the oversimplification of the model, where water molecules are classified into two groups with distinct demarcation. It does not consider the reduced dielectric constant in the structure breaking region.

2.42 Calculate the heat of interaction between the three individual ions, Cl⁻, Na⁺ and Br⁻, and water, i.e., $\Delta H_{Cl^- -w}$, $\Delta H_{Na^+ -w}$, $\Delta H_{Br^- -w}$. Use the following experimental values of the heats of interaction between a salt and water at 25 ⁰C:

Salt	KF	KCl	NaF	NaBr
ΔH_{salt-w} (kJ mol⁻¹)	-827.6	-685.3	-911.2	-741.8

Consider the Born model as valid. (Cf. Problem 2.5 in the textbook) (Constantinescu)

Answer:

The heat of interaction between individual ions and water is related to the heat of interaction between the salt by (cf. Eq. 2.32 in the textbook) $\Delta H_{salt-w} = \Delta H_{A^+ -w} + \Delta H_{X^- -w}$. The Born model involves the postulate that ions that have equal crystallographic radii also have equal interaction with the solvent. The pair KF is chosen to start the calculations because the radii of the K^+ and F^- ions are almost equal (see tables in Chapter 1). Thus,

$$\Delta H_{KF-w} = \Delta H_{K^+-w} + \Delta H_{F^--w} \qquad (2.152)$$

Since $r_{K^+} \approx r_{F^-}$, then,

$$\Delta H_{K^+-w} \approx \Delta H_{F^--w} = \frac{\Delta H_{KF-w}}{2} = \frac{-827.6 \text{ kJ mol}^{-1}}{2}$$
$$= -413.8 \text{ kJ mol}^{-1} \qquad (2.153)$$

Once these values are obtained, the other values can be calculated:

$$\Delta H_{KCl-w} = \Delta H_{K^+-w} + \Delta H_{Cl^--w} \qquad (2.154)$$

or

$$\Delta H_{Cl^--w} = \Delta H_{KCl-w} - \Delta H_{K^+-w} = -685.3 - (-413.8)$$
$$= -271.5 \text{ kJ mol}^{-1} \qquad (2.155)$$

In the same way,

$$\Delta H_{NaF-w} = \Delta H_{Na^+-w} + \Delta H_{F^--w} \qquad (2.156)$$

or

$$\Delta H_{Na^+-w} = \Delta H_{NaF-w} - \Delta H_{F^--w} = -911.2 - (-413.8)$$
$$= -497.4 \text{ kJ mol}^{-1} \qquad (2.157)$$

and,

$$\Delta H_{NaBr-w} = \Delta H_{Na^+-w} + \Delta H_{Br^--w} \qquad (2.158)$$

or

$$\Delta H_{Br^--w} = \Delta H_{NaBr-w} - \Delta H_{Na^+-w} = -741.8 - (-497.4)$$
$$= -244.4 \text{ kJ mol}^{-1} \qquad (2.159)$$

Comment. The differences between the Born values for heats of ion-solvent (heats of solvation) and these "experimental" values are numerically too high, in some cases nearly 50% too high. To reduce this difference between the theory and the experimental values, the ion-dipole and the ion-quadrupole theory were developed. The basis of these theories is knowledge of the structure of the solvent in the bulk and in the region around the ions.

2.43 (a) Write an expression for the potential in a point P of the electric field created by a dipole of moment μ. (b) Using the expression developed in (a), write an expression for the potential energy of ion-dipole interaction. (c) Calculate the potential energy, E_p, of ion-dipole interaction between water and a z-valent cation, as a function of the distance r, the ion charge z_i, the angle θ, and the relative permitivity ε. (d) Perform a complete calculation for the limiting cases, i.e., $z_i = 1$, $\theta = 0$, $r = 2$ Å, $\varepsilon = 4.5$, and $z_i = 1$, $\theta = 0$, $r = 6$ Å, $\varepsilon = 80$. Assume that the relative positions are as in the Fig. 2.6, with the negative end of the water dipole facing the positive ion. (e) Assuming the intermediate values of ε increase exponentially with r, draw a complete E_p vs. r curve for the interval $2 < r < 6$, and mark the region in which the thermal energy, $\sim RT$, is competitive with the electric interaction. (cf. Problem 2.25 in the textbook) (Mussini)

Answer:

(a) The potential generated by one q charge at a distance r in vacuum is given by

$$V = \frac{q}{4\pi\varepsilon_0 r} \tag{2.160}$$

and the potential generated by more q_i charges,

$$V = \frac{1}{4\pi\varepsilon_0} \sum_i \frac{q_i}{r_i} \tag{2.161}$$

Therefore, the potential generated by the dipole in Fig. 2.6(b) in a point P at a distance r from the center of the dipole is,

$$V = \frac{1}{4\pi\varepsilon_0}\left(\frac{q}{r_i} - \frac{q}{r_2}\right) = \frac{q}{4\pi\varepsilon_0}\frac{r_2 - r_1}{r_1 r_2} \tag{2.162}$$

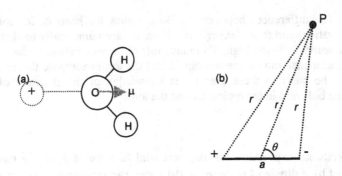

Figure 2.6. Relative positions of the atoms in a water molecule according to Problem 2.43.

Now, if $a \ll r$ (i.e., the point P is far away from the dipole), then $r_2 - r_1 \cong a \cos \theta$, and $r_2 r_1 \cong r^2$, and

$$V \approx \frac{q}{4 \pi \varepsilon_0} \frac{a \cos \theta}{r^2} \approx \frac{\mu \cos \theta}{4 \pi \varepsilon_0 r^2} \qquad (2.163)$$

(b) The potential energy in vacuum is given by,

$$E_p = Vq = \frac{\mu q \cos \theta}{4 \pi \varepsilon_0 r^2} \qquad (2.164)$$

and in a medium of relative permittivity ε,

$$E_p = \frac{Vq}{\varepsilon} = \frac{\mu q \cos \theta}{4 \pi \varepsilon_0 \varepsilon r^2} \qquad (2.165)$$

(c) Thus, using Eq. (2.165) to calculate E_p as a function of r in $\overset{\circ}{A}$, z_i, θ, and ε,

$$E_p = \frac{\mu z e_0 N_A \cos \theta}{4 \pi \varepsilon_0 \varepsilon r^2} = \frac{(1.86\,\mathrm{D})z \left(1.602 \times 10^{-19}\,\mathrm{C}\right)\left(6.022 \times 10^{23}\,\mathrm{mol}^{-1}\right)}{\left(1.112 \times 10^{-10}\,\mathrm{C}^2\mathrm{J}^{-1}\mathrm{m}^{-1}\right)\varepsilon_r \left(r^2\,\overset{\circ}{A}^2\right)} \times$$

$$\times \cos\theta \times \frac{3.336\times10^{-30}\ \text{C m}}{1\text{D}} \times \frac{\left(10^{10}\ \overset{0}{\text{A}}\right)^2}{1\text{m}^2} = \frac{538.405\,z\cos\theta}{\varepsilon_r\left[r\left(\overset{0}{\text{A}}\right)\right]^2} \tag{2.166}$$

where E_p is given in kJ mol^{-1}.

(d) For $z_i = 1$, $\theta = 0^0$, $r = 2$ Å, and $\varepsilon = 4.5$,

$$E_p = \frac{538.405(1)\cos 0^0}{4.5(2)^2} = 29.9\ \text{kJ mol}^{-1} \tag{2.167}$$

and for $z_i = 1$, $r = 6$ Å, and $\varepsilon = 80$,

$$E_p = \frac{538.405(1)\cos 0^0}{80(6)^2} = 0.2\ \text{kJ mol}^{-1} \tag{2.168}$$

(e) If ε increases exponentially with r, then the equation that satisfies this condition is

$$\varepsilon = Ae^{Br} \tag{2.169}$$

To evaluate the constants A and B, we make use of the conditions that at $r = 2$ Å, $\varepsilon = 4.5$, and that at $r = 6$ Å, $\varepsilon = 80$. Thus, from the two equations $\ln 4.5 = \ln A + 2B$ and $\ln 80 = \ln A + 6B$, the constants $A = 1.067$, and $B = 0.719$. Substituting these values into Eq. (2.169),

$$\varepsilon = 1.067\,e^{0.719r} \tag{2.170}$$

Therefore, the corresponding equation of E_p is:

$$E_p = \frac{504.600z\cos\theta}{r^2\,e^{0.719r}} \quad \text{in}\quad \text{kJ mol}^{-1} \tag{2.171}$$

where r is given in Å. A plot of this equation is given in Fig. 2.7. When $z = 1$ and $\theta = 0^0$.

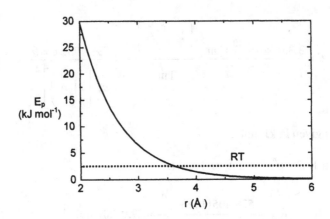

Figure 2.7. Potential energy due to the dipole in Fig. 2.6 as a function of the distance r.

The value of $RT = (8.314 \text{ J mol}^{-1} \text{ K}^{-1})(298 \text{ K}) = 2.481 \text{ kJ mol}^{-1}$ is marked by a broken line in Fig. 2.7. The thermal energy, RT, becomes competitive when the distance between the point P and the dipole is larger than 3.5 Å.

2.44 (a) Calculate the heat of ion-solvent interaction considering only the Born term, i.e., ΔH_{BC}, for Cl⁻ ions at 298 K. The dielectric constants of water at three different temperatures are:

Temperature (^0C)	20	25	30
ε	80.1	78.3	76.54

(b) Considering that chloride ions are surrounded by four water molecules in the gas phase calculate their ion-dipole interaction energy. (c) Calculate $\Delta H_{i\text{-}w}$ for chloride ions using the ion-dipole model at 298 K. (Cf. Problem 2.8 in the textbook) (Contractor)

Answer:

(a) The Born term for ion-solvent interaction is (cf. Eq. 2.42 in textbook):

$$\Delta H_{BC} = -\frac{N_A (z_i e_o)^2}{2(r_i + 2r_W) 4\pi\varepsilon_0} \left[1 - \frac{1}{\varepsilon} - \frac{T}{\varepsilon^2} \left(\frac{\partial\varepsilon}{\partial T} \right)_P \right] \qquad (2.172)$$

To estimate the derivative $(\partial\varepsilon/\partial T)_P$ we use the central difference formula [cf. Numerical Methods for Mathematics, Science and Engineering, by J. H. Mathews, Prentice-Hall, 2nd Ed., (1994), p. 318]:

$$\frac{\partial\varepsilon}{\partial T}\bigg|_{25C} = \frac{\varepsilon_{T_1} - \varepsilon_{T_2}}{T_1 - T_2} = \frac{76.54 - 80.1}{30C - 20C} = -0.356\,K^{-1} \qquad (2.173)$$

Substituting values for the Cl⁻-water interaction into Eq. (2.172):

$$\Delta H_{BC}(Cl^- - w) = -\frac{6.023 \times 10^{23}\ mol^{-1} (-1)^2 \left(1.602 \times 10^{-19}\ C\right)^2}{2\left[181 \times 10^{-12}\ m + (2)138 \times 10^{-12}\ m\right]\left(1.112 \times 10^{-10}\ C^2 J^{-1} m^{-1}\right)}$$

$$\times \left[1 - \frac{1}{78.3} - \frac{298\ K}{78.3^2} \left(0.356\ K^{-1} \right) \right] = -147.5\ kJ\ mol^{-1} \qquad (2.174)$$

(b) From the equation (cf. Eq. 2.41 in the textbook):

$$W_{i-D} = -\frac{nN_A z_i e_o \mu_W}{(r_i + r_W)^2\, 4\pi\varepsilon_0} \qquad (2.175)$$

$$= -\frac{4\left(6.023 \times 10^{23}\ mol^{-1}\right)\left|-1\right|\left(1.602 \times 10^{-19}\ C\right)1.8\,D}{\left(181 \times 10^{-12}\ m + 138 \times 10^{-12}\ m\right)^2\left(1.112 \times 10^{-10}\ C^2 J^{-1} m^{-1}\right)}$$

$$\times \frac{3.336 \times 10^{-30}\ Cm}{1D} = -204.8\ kJ\ mol^{-1}$$

(c) The enthalpy of ion-water interaction for negative ions is (cf. Eq. 2.48 in the textbook):

$$\Delta H_{i-w} = 30\,kcal\,mol^{-1} \qquad (2.176)$$

$$-\frac{4N_A z_i e_o \mu_W}{(r_i + r_W)^2\, 4\pi\varepsilon_0} - \frac{N_A (z_i e_o)^2}{2(r_i + 2r_W) 4\pi\varepsilon_0} \left[1 - \frac{1}{\varepsilon} - \frac{T}{\varepsilon^2} \left(\frac{\partial\varepsilon}{\partial T} \right)_P \right]$$

Substituting the corresponding parameters, and making use of Eq. (2.174),

$$\Delta H_{Cl^- -w} = \left(30 \text{ kcal mol}^{-1}\right) \frac{4.186 \text{ kJ}}{1 \text{ kcal}} - \frac{4\left(6.023 \times 10^{23} \text{ mol}^{-1}\right)|-1|}{\left(195 \times 10^{-12} \text{ m} + 138 \times 10^{-12} \text{ m}\right)^2}$$

$$\times \frac{\left(1.602 \times 10^{-19} \text{ C}\right)1.8 \text{ D}}{\left(1.112 \times 10^{-10} \text{ C}^2 \text{ J}^{-1} \text{ m}^{-1}\right)} \times \frac{3.336 \times 10^{-30} \text{ C m}}{1 \text{ D}} \frac{1 \text{ kJ}}{1000 \text{ J}}$$

$$- 147.5 \text{ kJ mol}^{-1} = -209.8 \text{ kJ mol}^{-1} \tag{2.177}$$

2.45 Estimate the error introduced in calculating the heat of interaction between NaBr and water when the distortion of the charge distribution of water molecules due to the electrical field is ignored. Consider the values of dipole moment, quadrupole moment, and polarizability of water as indicated in the Table section. The dielectric constant of water at 25 ^{0}C is ε_w = 78.5, and $\left(\partial \varepsilon_w / \partial T\right)$= -0.356 K^{-1}. (Cf. Problem 2.9 in the textbook) (Contractor)

Answer:

For positive ions (cf. Eq. 2.52 in the textbook) the heat of ion-water interaction is:

$$\Delta H_{i-w} = 20 - \frac{4 N_A z_i e_0 \mu_W}{(r_i + r_W)^2 \, 4 \pi \varepsilon_0} + \frac{4 N_A z_i e_0 p_W}{2(r_i + r_W)^3 \, 4 \pi \varepsilon_0} \tag{2.178}$$

$$- \frac{N_A (z_i e_0)^2}{2(r_i + 2 r_W) 4 \pi \varepsilon_0} \left[1 - \frac{1}{\varepsilon} - \frac{T}{\varepsilon^2}\left(\frac{\partial \varepsilon}{\partial T}\right)_P\right] - \frac{4 N_A \alpha (z_i e_0)^2}{2(r_i + r_w)^4 \, 4 \pi \varepsilon_0}$$

Evaluating this equation for Na$^+$ ions,

$$\Delta H_{Na^+ -w} = \left(20 \text{ kcal mol}^{-1}\right) \frac{4.186 \text{ kJ}}{1 \text{ kcal}} \tag{2.179}$$

$$- \frac{4\left(6.023 \times 10^{23} \text{ mol}^{-1}\right)(+1)\left(1.602 \times 10^{-19} \text{ C}\right)1.8 \text{ D}}{\left(95 \times 10^{-12} \text{ m} + 138 \times 10^{-12} \text{ m}\right)^2 \left(1.112 \times 10^{-10} \text{ C}^2 \text{ J}^{-1} \text{ m}^{-1}\right)} \frac{3.336 \times 10^{-30} \text{ C m}}{1 \text{ D}} +$$

$$+\frac{4\left(6.023\times10^{23}\ \text{mol}^{-1}\right)(+1)\left(1.602\times10^{-19}\ \text{C}\right)3.9\times10^{-10}\ \text{D m}}{2\left(95\times10^{-12}\ \text{m}+138\times10^{-12}\ \text{m}\right)^{3}\left(1.112\times10^{-10}\ \text{C}^{2}\text{J}^{-1}\text{m}^{-1}\right)}\ \frac{3.336\times10^{-30}\ \text{C m}}{1\ \text{D}}$$

$$-\frac{6.023\times10^{23}\ \text{mol}^{-1}\ (+1)^{2}\left(1.602\times10^{-19}\ \text{C}\right)^{2}}{2\left[95\times10^{-12}\ \text{m}+(2)138\times10^{-12}\ \text{m}\right]\left(1.112\times10^{-10}\ \text{C}^{2}\text{J}^{-1}\text{m}^{-1}\right)}$$

$$\times\left[1-\frac{1}{78.3}-\frac{298\ \text{K}}{78.3^{2}}\left(0.356\ \text{K}^{-1}\right)\right]$$

$$-\frac{4\left(6.023\times10^{23}\ \text{mol}^{-1}\right)\left(1.46\times10^{-24}\ \text{cm}^{3}\right)(+1)^{2}\left(1.602\times10^{-19}\ \text{C}\right)^{2}}{2\left(95\times10^{-12}\ \text{m}+138\times10^{-12}\ \text{m}\right)^{4}\left(1.112\times10^{-10}\ \text{C}^{2}\text{J}^{-1}\text{m}^{-1}\right)}\ \frac{1\ \text{m}^{3}}{10^{6}\ \text{cm}^{3}}$$

$$=-441.11\ \text{kJ mol}^{-1}$$

For negative ions (cf. Eq. 2.53 in textbook) the heat of ion-water interaction is:

$$\Delta H_{i-w}=30-\frac{4N_{A}z_{i}e_{o}\mu_{W}}{(r_{i}+r_{W})^{2}\ 4\pi\varepsilon_{0}}-\frac{4N_{A}z_{i}e_{o}p_{W}}{2(r_{i}+r_{W})^{3}\ 4\pi\varepsilon_{0}} \tag{2.180}$$

$$-\frac{N_{A}(z_{i}e_{o})^{2}}{2(r_{i}+2r_{W})4\pi\varepsilon_{0}}\left[1-\frac{1}{\varepsilon}-\frac{T}{\varepsilon^{2}}\left(\frac{\partial\varepsilon}{\partial T}\right)_{P}\right]-\frac{4N_{A}\alpha(z_{i}e_{0})^{2}}{2(r_{i}+r_{w})^{4}\ 4\pi\varepsilon_{0}}$$

Therefore,

$$\Delta H_{Br^{-}-w}=\left(30\ \text{kcal mol}^{-1}\right)\frac{4.186\ \text{kJ}}{1\ \text{kcal}}$$

$$-\frac{4\left(6.023\times10^{23}\ \text{mol}^{-1}\right)|-1|\left(1.602\times10^{-19}\ \text{C}\right)1.8\ \text{D}}{\left(195\times10^{-12}\ \text{m}+138\times10^{-12}\ \text{m}\right)^{2}\left(1.112\times10^{-10}\ \text{C}^{2}\text{J}^{-1}\text{m}^{-1}\right)}\ \frac{3.336\times10^{-30}\ \text{Cm}}{1\ \text{D}}$$

$$-\frac{4\left(6.023\times10^{23}\ \text{mol}^{-1}\right)|-1|\left(1.602\times10^{-19}\ \text{C}\right)3.9\times10^{-10}\ \text{D m}}{2\left(195\times10^{-12}\ \text{m}+138\times10^{-12}\ \text{m}\right)^{3}\left(1.112\times10^{-10}\ \text{C}^{2}\text{J}^{-1}\text{m}^{-1}\right)}\ \frac{3.336\times10^{-30}\ \text{Cm}}{1\ \text{D}}$$

$$\frac{6.023 \times 10^{23} \text{ mol}^{-1} (-1)^2 \left(1.602 \times 10^{-19} \text{ C}\right)^2}{2\left[195 \times 10^{-12} \text{ m} + (2)138 \times 10^{-12} \text{ m}\right]\left(1.112 \times 10^{-10} \text{ C}^2 \text{ J}^{-1} \text{ m}^{-1}\right)}$$

$$\times \left[1 - \frac{1}{78.3} - \frac{298 \text{K}}{78.3^2}\left(0.356 \text{K}^{-1}\right)\right] \tag{2.181}$$

$$\frac{4\left(6.023 \times 10^{23} \text{ mol}^{-1}\right)\left(1.46 \times 10^{-24} \text{ cm}^3\right)(-1)^2 \left(1.602 \times 10^{-19} \text{ C}\right)^2}{2\left(195 \times 10^{-12} \text{ m} + 138 \times 10^{-12} \text{ m}\right)^4 \left(1.112 \times 10^{-10} \text{ C}^2 \text{ J}^{-1} \text{ m}^{-1}\right)} \frac{1 \text{m}^3}{10^6 \text{ cm}^3}$$

$$= -300.00 \text{ kJ mol}^{-1}$$

Therefore, the total heat of NaBr-water interaction is,

$$\Delta H_{NaBr-w} = \Delta H_{Na^+ -w} + \Delta H_{Br^- -w} = -441.11 \text{kJ mol}^{-1} - 300.00 \text{kJ mol}^{-1}$$

$$= -741.09 \text{ kJ mol}^{-1} \tag{2.182}$$

The energy due to ion-induced dipole interactions involves the term $-\dfrac{4N_A \alpha (z_i e_0)^2}{2(r_i + r_w)^4 \, 4\pi\varepsilon_0}$. Evaluating this term for Na^+ and Br^-, the total ion-induced dipole interaction is 137.72 kJ mol^{-1} + 33.01 kJ mol^{-1} = -170.73 kJ mol^{-1}. Thus, the error in ΔH_{NaBr-w} committed by ignoring these terms is:

$$\frac{170.73}{741.09} \times 100 = 23.04\% \tag{2.183}$$

2.46 In a cation-solvent interaction model, cf. Fig. 2.37 in textbook, the "solvated-coordinated water" has two sites capable of forming hydrogen bonds with water molecules in the *SB* region. Are these two sites identical in bonding? For the "non-solvated coordinated water" there are three sites for hydrogen bonds. Are these three sites identical? Explain. (Cf. Problem 2.18 in the textbook) (Xu)

Answer:

The two sites in the solvated coordinated water are both acceptor in nature. On the other hand, in the non-solvated coordinated water one of the sites is electron donor in nature since it is on one of the lone pairs of oxygen that is disoriented from the ion.

2.47 An ion of charge ze_0 and radius r is transferred from a solvent of dielectric constant ε_i to a solvent of dielectric constant ε_f. Derive an expression to calculate the free energy change associated with this transfer using the Born model. (cf. Problem 2.14 in the textbook) (Contractor).

Answer:

The process of transferring an ion from one solvent molecule to another solvent molecule, is equivalent to subtract the individual transfer of the ion from vacuum to each one of the solvents. Thus,

$$\Delta G_{i \to f} = G_f - G_i = \left(G_f - G_{vacuum} \right) - \left(G_i - G_{vacuum} \right)$$
$$= \Delta G_{vacuum \to f} - \Delta G_{vacuum \to i} \tag{2.184}$$

Therefore, from Born equation (cf. Eq. A.2.1.6 in the textbook),

$$\Delta G_{i \to f} = \left[-\frac{N_A (ze_0)^2}{2r} \left(1 - \frac{1}{\varepsilon_f} \right) \right] - \left[-\frac{N_A (ze_0)^2}{2r} \left(1 - \frac{1}{\varepsilon_i} \right) \right]$$
$$= \frac{N_A (ze_0)^2}{2r} \left(\frac{1}{\varepsilon_f} - \frac{1}{\varepsilon_i} \right) \tag{2.185}$$

2.48 What is the free energy change involved in transferring chloride ions from water to a nonpolar medium like carbon tetrachloride with a dielectric constant of 2.23 at 25 ^{0}C? Is this an energetically favorable process? Comment on your answer. Make use of the equation developed in Problem 2.47. (cf. Problem 2.15 in the textbook) (Contractor)

Answer:

Using the equation developed in Problem 2.47:

$$\Delta G_{i \to f} = \frac{N_A (ze_0)^2}{2r} \left(\frac{1}{\varepsilon_f} - \frac{1}{\varepsilon_i} \right) \tag{2.186}$$

which is valid when the charge is given in electrostatic units, the distance in centimeters and the free energy in ergs per mol. Introducing the term $4\pi\varepsilon_0$ in the denominator changes the charge units to coulombs, the distance to meters, and the free energy to joules per mol:

$$\Delta G_{i\to f} = \frac{N_A (ze_0)^2}{4\pi\varepsilon_0 \, 2r}\left(\frac{1}{\varepsilon_f} - \frac{1}{\varepsilon_i}\right) \tag{2.187}$$

Substituting values into this equation:

$$\Delta G_{i\to f} = \frac{6..023\times10^{23}\,\text{mol}^{-1}(-1)^2\left(1.602\times10^{-19}\,\text{C}\right)^2}{\left(1.112\times10^{-10}\,\text{C}^2\,\text{J}^{-1}\,\text{m}^{-1}\right)(2)\left(1.81\times10^{-10}\,\text{m}\right)} \tag{2.188}$$

$$\times\left(\frac{1}{2.23} - \frac{1}{78.54}\right) = +167.22\,\text{kJ mol}^{-1}$$

Since $\Delta G_{i\to f}$ is a large positive number, the process is not favorable. This result can be understood qualitatively even without involving the Born model. Water has a large dipole moment and therefore the ion-dipole interactions are quite strong in this solvent. The opposite occurs in carbon tetrachloride with low dielectric constant. Therefore, the ions would prefer to be in water.

2.49 Living cells are surrounded by bilayered membranes, and aqueous environments are present on both sides of the membrane. The interior of the membrane is highly non-polar in nature. Based on the result of Problem 2.48, explain why the transportation of charges across a bilayered membrane in a living cell is difficult. (Cf. Problem 2.16 in the textbook) (Contractor)

Answer:

On either side of the membrane, which is non-polar in nature, aqueous environment is present.

Transporting a charged ion across a membrane involves two processes (see Fig 2.8): (1) transporting the ion from the first aqueous environment to the non-polar membrane medium, and (2) transporting the ion from the non-polar membrane to the other side where a second aqueous medium exists. According to Problem 2.48, process (1) is not a favorable process, though process (2) is a

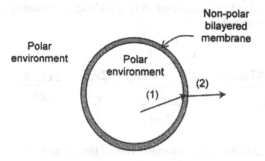

Figure 2.8. A bilayered membrane

favorable one. Thus, it is difficult to transport the charged ions across the membrane. To help this two-step process, living cells use protein pumps that are located across the membrane to transport charged ions through them.

2.50 Estimate the error introduced by ignoring the size of the solvent molecules in calculating the heat of the Born-charging process during the interaction of a Cs$^+$ ion with water. (Contractor)

Answer:

The Born-charging contribution (cf. Eq. A.2.1.6 in the textbook) taking into account the size of the solvent molecule is given by

$$\Delta G_{Born,1} = -\frac{N_A \left(z_i e_0\right)^2}{2\left(r_i + 2r_w\right)}\left[1 - \frac{1}{\varepsilon_s}\right] \tag{2.189}$$

At constant temperature, the above equation can be written as,

$$\Delta G_{Born,1} = \frac{K}{\left(r_i + 2r_w\right)} \tag{2.190}$$

In the same way, if the size of the solvent is ignored, then,

$$\Delta G_{Born,2} = \frac{K}{\left(r_i\right)} \tag{2.191}$$

The error involved in calculating ΔG_{Born} without considering the size of the solvent can be calculated as:

$$\frac{\Delta G_{Born,2} - \Delta G_{Born,1}}{\Delta G_{Born,1}} = \frac{\dfrac{K}{r_i} - \dfrac{K}{r_i + 2r_w}}{\dfrac{K}{r_i + 2r_w}} = \frac{2r_w}{r_i} = \frac{2 \times 1.38}{1.69} = 1.63 \quad (2.192)$$

That is, ΔG_{Born} is **63%** larger when the size of the solvent is ignored.

2.51 (a) Calculate the volumes of a bare ion and a solvated ion sheathed by water. Make the calculation for Li$^+$, Na$^+$ and K$^+$. Use the corresponding ionic radii. (b) If the structure-breaking region consists of two layers of water molecules, calculate the volume of water that has been affected by a single ion. (Cf. Problem 2.19 in the textbook) (Xu)

Answer:

(a) The volume of the *bare* ion is calculated as:

$$V_i = \frac{4}{3}\pi r_i^3 \quad (2.193)$$

To calculate the volume of the solvated ion, it is assumed that one layer of water molecules (SN) constitute the solvation sheath. Thus,

$$V_{i+SN} = \frac{4}{3}\pi\left(r_i + 2r_w\right)^3 \quad (2.194)$$

For Li$^+$ ion,

$$V_{Li^+} = \frac{4}{3}\pi\left(0.59 \times 10^{-8}\ \text{cm}\right)^3 = 0.86 \times 10^{-24}\ \text{cm}^3 \quad (2.195)$$

and Eq. (2.194) becomes,

$$V_{Li^+ + SN} = \frac{4}{3}\pi\left(0.59 \times 10^{-8} + 2 \times 1.38 \times 10^{-8}\right)^3 = 1.57 \times 10^{-22}\ \text{cm}^3 \quad (2.196)$$

The value of V_i and V_{i+SN} for the other ions are given in the table at the end of this problem.

b) The volume of water affected by an ion constitutes the volume of solvation waters and structure breaking region, i.e., V_{SN+SB}. Thus,

$$V_{i+SN+SB} = V_i + V_{SN+SB} \tag{2.197}$$

or

$$V_{SN+SB} = V_{i+SN+SB} - V_i \tag{2.198}$$

Since the SB region consists of two layers of water molecules,

$$
\begin{aligned}
V_{SN+SB} &= \frac{4}{3}\pi\left(r_i + 2r_{w-SN} + 4r_{w-SB} \right)^3 - \frac{4}{3}\pi r_i^3 \\
&= \frac{4}{3}\pi\left[\left(r_i + 2r_{w-SN} + 4r_{w-SB} \right)^3 - r_i^3 \right]
\end{aligned} \tag{2.199}
$$

For Li$^+$,

$$
V_{SN+SB} = \frac{4}{3}\pi\left\{ \left[0.59 \times 10^{-8} + 6\left(1.38 \times 10^{-8} \right) \right]^3 - \left(0.59 \times 10^{-8} \right)^3 \right\} \tag{2.200}
$$
$$
= 2.92 \times 10^{-21} \text{ cm}^3
$$

The values of V_i, V_{i+SN} and V_{SN+SB} for the three ions are listed in the following table:

	Li$^+$	Na$^+$	K$^+$
V_i (cm^3)	0.86 x 10^{-24}	3.59 x 10^{-24}	9.85 x 10^{-24}
V_{i+SN} (cm^3)	1.57 x 10^{-22}	2.14 x 10^{-22}	2.86 x 10^{-22}
V_{SN+SB} (cm^3)	2.92 x 10^{-21}	3.36 x 10^{-21}	3.76 x 10^{-21}

2.52 Using the results of Problem 2.51, calculate the percentage of *bulk* water in 0.05 M NaCl solution. What is the percentage if the concentration is 0.5 M? Comment on the significance of the results. (Cf. Problem 2.20 in the textbook) (Xu)

Answer:

Using the results of Problem 2.51(b), the volume of water affected by a single Cl⁻ ion is determined from Eq. (2.199),

$$V_{SN+SB} = \frac{4}{3}\pi\left[\left(r_i + 2r_{w-SN} + 4r_{w-SB}\right)^3 - r_i^3\right] \tag{2.201}$$

$$= \frac{4}{3}\pi\left\{\left[1.81 \times 10^{-8} \text{ cm} + 6\left(1.38 \times 10^{-8} \text{ cm}\right)\right]^3 - \left(1.38 \times 10^{-8} \text{ cm}\right)^3\right\}$$

$$= 4.29 \times 10^{-21} \text{ cm}^3$$

Therefore, in 1 dm^3 of 0.05 M NaCl solution, the total volume of water affected by Cl⁻ and Na⁺ ion is:

$$\left(V_{SN+SB}\right)_{Total} = 0.05 N_A \left[\left(V_{SN+SB}\right)_{Na^+} + \left(V_{SN+SB}\right)_{Cl^-}\right]$$

$$= \left(0.05 \text{ mol}\right)\left(6.022 \times 10^{23} \text{ mol}^{-1}\right)\left(3.36 \times 10^{-21} \text{ cm}^3 + 4.29 \times 10^{-21} \text{ cm}^3\right)$$

$$= 230.38 \text{ cm}^3 \quad \text{or} \quad 0.23 \text{ dm}^3 \tag{2.202}$$

This indicates that $\frac{0.23 \text{ dm}^3}{1 \text{ dm}^3} \times 100 = 23\%$ is the percentage of water volume affected by the ions and **77%** is the volume of free water. If the concentration would increase 10 times, to 0.5 M, the above method predicts that there would not be more free water. This warns that in concentrations above 1.0 M the rules applicable in dilute solutions no longer hold true because the bulk water has practically disappeared.

2.53 Consider a dilute gas of polar molecules (permanent dipoles) in an electric field generated from the charge on the plates of a condenser [Figs. 2.9(a) and 2.9(b)]. Different forces affect the alignment of the molecules (a) What are these forces? (b) The number of dipoles per unit solid angle at an angle θ to the applied field, n_θ, [Fig. 2.9(c)] is given by the Boltzmann distribution law:

$$n_\theta = \text{Re}^{-W/kT} \quad \text{and} \quad dn_\theta = \text{Re}^{-W/kT} d\Omega \tag{2.203}$$

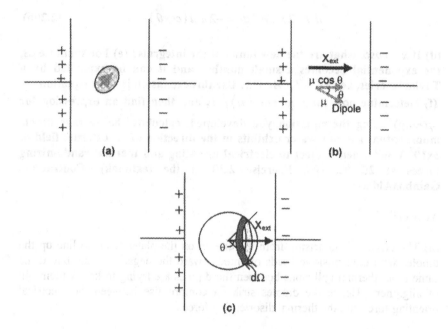

Figure 2.9. Permanent dipoles in an electric field of a condenser.

where R is a proportionality constant, W is the work done by the molecule in aligning with the field

$$(W = -\mu \overrightarrow{X_{ext}} \cos\theta)$$ (2.204)

and Ω is the solid angle to the external field [Fig. 2.9(c)]. Consider the dipoles oriented at angles θ_1, θ_2,... The dipole moments in the direction of the field are only $\mu\cos\theta_1$, $\mu\cos\theta_2$, ... [Fig. 2.9(b)]. Thus, the *average* moment $<\mu>$ is defined as

$$\langle\mu\rangle = \frac{\int_0^\pi \mu\cos\theta \, dn_\theta}{\int_0^\pi dn_\theta}$$ (2.205)

Substitute in this equation the equivalent of dn_θ and W. (c) Substitute now

$$d\Omega = 2\pi \sin\theta \, d\theta = -2\pi \, d(\cos\theta) \qquad (2.206)$$

(d) If $x = \cos\theta$, what are the new limits of the integrals? (e) For small fields, the exponential becomes a small number and it can be expanded by a Taylor's series, i.e., $e^{Ay} = 1 + Ay + \dots$ Use this expansion in your equation.

(f) Determine now the quotient $\langle \mu \rangle / \mu$, and then find an expression for $<\mu>$. (g) Using the equation you developed, calculate the average dipole moment that gaseous water exhibits in the direction of an external field of 3×10^7 V m^{-1} when subject to electrical orienting and thermal randomizing forces at 25 ^{0}C. (Cf. Exercise 2.30 in the textbook) (Contractor-GamboaAldeco)

Answer:

(a) The electric field arising from the charge on the plates tends to line up the dipoles with their positive heads oriented toward the negative plate, but at the same time, thermal collisions between the dipoles are trying to knock them out of alignment. Hence the dipoles strike a compromise between the electrical orienting force and the thermal disorienting force.

(b) Substituting Eqs. (2.203) and (2.204) into Eq. (2.205)

$$\langle \mu \rangle = \frac{\int_0^\pi \mu(\cos\theta) Re^{(\cos\theta)\mu \vec{X}_{ext}/kT} \, d\Omega}{\int_0^\pi Re^{(\cos\theta)\mu \vec{X}_{ext}/kT} \, d\Omega} \qquad (2.207)$$

(c) Substituting now Eq. (2.206) into Eq. (2.207) and simplifying,

$$\langle \mu \rangle = \frac{\int_0^\pi \mu(\cos\theta)(-2\pi) Re^{(\cos\theta)\mu \vec{X}_{ext}/kT} \, d\cos\theta}{\int_0^\pi (-2\pi) Re^{(\cos\theta)\mu \vec{X}_{ext}/kT} \, d\cos\theta}$$

$$= \frac{\int_\pi^0 \mu(\cos\theta) e^{(\cos\theta)\mu \vec{X}_{ext}/kT} \, d\cos\theta}{\int_\pi^0 e^{(\cos\theta)\mu \vec{X}_{ext}/kT} \, d\cos\theta} \qquad (2.208)$$

(d) If $x = \cos\theta$, then, when $\theta = 0 \Rightarrow x = 1$, and when $\theta = \pi \Rightarrow x = -1$. Therefore, Eq. (2.208) becomes

$$\langle \mu \rangle = \frac{\int_{-1}^{+1} \mu x e^{x\mu \vec{X}_{ext}/kT} dx}{\int_{-1}^{+1} e^{x\mu \vec{X}_{ext}/kT} dx} \tag{2.209}$$

(e) For small fields, $x\mu \vec{X}_{ext}/kT \ll 1$, and $e^{x\mu \vec{X}_{ext}/kT} \approx 1 + x\mu \vec{X}_{ext}/kT$. Thus, Eq. (2.209) becomes

$$\langle \mu \rangle = \frac{\int_{-1}^{+1} \mu x \left(1 + \frac{x\mu \vec{X}_{ext}}{kT}\right) dx}{\int_{-1}^{+1} \left(1 + \frac{x\mu \vec{X}_{ext}}{kT}\right) dx} \tag{2.210}$$

(f) The quotient $\langle \mu \rangle / \mu$ is obtained by dividing Eq. (2.210) by μ:

$$\frac{\langle \mu \rangle}{\mu} = \frac{\int_{-1}^{+1} x\,dx + \int_{-1}^{+1} \frac{x^2 \mu \vec{X}_{ext}}{kT} dx}{\int_{-1}^{+1} dx + \int_{-1}^{+1} \frac{x\mu \vec{X}_{ext}}{kT} dx} = \frac{\mu \vec{X}_{ext}}{3kT} \tag{2.211}$$

or

$$\langle \mu \rangle = \frac{\mu^2 \vec{X}_{ext}}{3kT} \tag{2.212}$$

(g) Finally, the average dipole moment of gaseous water is, from Eq. (2.212),

$$\langle \mu \rangle = \frac{(1.87D)^2 \left(3 \times 10^7 \text{ V m}^{-1}\right)}{3\left(1.381 \times 10^{-23} \text{ J K}^{-1}\ 298\text{K}\right)} \frac{1\text{J V}^{-1}}{1\text{C}} \frac{3.336 \times 10^{-30} \text{ C m}}{D} \tag{2.213}$$

$$= 0.028D$$

2.54 In Problem 2.53, the average dipole moment of a gas-phase dipole subjected to electrical and thermal forces was determined. This treatment can be applied satisfactory to a gas of water molecules that are not involved in mutual interactions. However, the dipole moment of a network structure such as *liquid* water cannot be determined through the same equation. Water is quasi-crystalline in the sense that there are in liquid water large groups ("icebergs") of water molecules associated by hydrogen bonding. A structural unit may be distinguished, that consists of a central water molecule tetrahedrically linked to four other molecules by hydrogen bonds (Fig. 2.10). When such a structure is placed in an electrical field, the whole subgroup aligns. Now, not the dipole moment of isolated molecules matters, but the dipole moment of the subgroups, μ_{group}. The effective moment of the group as a whole is equal to the dipole moment of the central molecule plus the components of the dipole moments of the four neighboring water molecules of the tetrahedral unit. In other words, the effective moment is the vector sum of the dipoles in the group, i.e.,

$$\mu_{group} = \mu + g\left(\mu\,\overline{cos\gamma}\right) = \mu\left(1 + g\,\overline{cos\gamma}\right) \tag{2.214}$$

where g is the number of nearest-neighbor water molecules linked with the central molecule and $\overline{cos\gamma}$ is the average of the cosines of the angles between the dipole moment of the central water molecule and those of its bonded neighbors. Finally, it is possible to apply the same arguments applied before to the gaseous molecules and use the equation developed in Problem 2.53 (i.e., Eq. 2.212), to obtain:

$$\left\langle \mu_{group} \right\rangle = \frac{\mu^2\left(1 + g\,\overline{cos\gamma}\right)^2}{3kT}\vec{X} \tag{2.215}$$

where $\vec{X}$ is the *total* field that operates on the group, and is given by

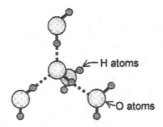

Figure 2.10. "Icebergs" of water molecules.

$$\vec{X} = \frac{3\varepsilon}{2\varepsilon+1}\vec{X}_{ext} \qquad (2.216)$$

If the average $\overline{cos\gamma} = 1/3$, and the tetrahedral clusters are subjected to an external field of 2.7×10^8 V cm^{-1}, estimate the effective moment of a dipole cluster of water at 298 K. (Cf. Exercise 2.31 in the textbook) (Contractor-GamboaAldeco)

Answer:

Substituting values in Eq. (2.215),

$$\langle \mu_{group} \rangle = \frac{(1.87D)^2 \left(1+4\left(\frac{1}{3}\right)\right)^2}{3\left(1.381 \times 10^{-23} \ JK^{-1} \ 298 K\right)} \frac{3(78.5)}{2(78.5+1)} \left(2.7 \times 10^8 \ V \ cm^{-1}\right)$$

$$\times \frac{1 JV^{-1}}{1C} \frac{3.336 \times 10^{-30} \ C \ m}{D} \frac{100 \ cm}{1 \ m} = 205.8 D \qquad (2.217)$$

2.55 Consider the interface between the metallic plates of a capacitor (with charge density q_M) and the dielectric material aligned in between the two plates (with a charge density q_{dipole} on each extreme) (See Fig. 2.11). (a) Using Gauss's law, prove that the net field that is set up in the dielectric material as a result of the external field and the internal counterfield is

$$\vec{X}_{ext} = \frac{q_M - q_{dipole}}{\varepsilon_0} \qquad (2.218)$$

In this equation, the charge q_{dipole} can be approximated as

$$q_{dipole} = 4 \pi \varepsilon_0 n \alpha \vec{X}_{ext} \qquad (2.219)$$

where α is the polarizability of the molecules (susceptibility of the molecules to deform into dipoles). (b) Considering that $\vec{X}_{ext} = V/d$, and that the capacity is $C = q_M / V = \varepsilon\varepsilon_0 / d$, find a relationship between the dielectric constant as a function of the polarizability of the molecules. Now, the term q_{dipole} should include both, the permanent as well as the induced dipoles (those produced by the distortion of molecules subject to an electrical field). Therefore, $q_{dipole} = q_{perm} + q_{ind}$. The first term is given by

$$q_{perm} = n \left\langle \mu_{group} \right\rangle \tag{2.220}$$

where

$$\left\langle \mu_{group} \right\rangle = \frac{\mu^2 \left(1 + g\overline{\cos\gamma}\right)^2}{3kT} \vec{X} \tag{2.221}$$

(see Problem 2.54), and

$$\vec{X} = \frac{3\varepsilon}{2\varepsilon + 1} \vec{X}_{ext} \tag{2.222}$$

The second term is

$$q_{ind} = 4\pi\varepsilon_0 n\alpha_{ind} \vec{X} \tag{2.223}$$

(c) Find an expression for the dielectric constant in terms of α_{ind} and μ (d) Calculate the deformation polarizability of water at 25 ^{0}C if there are 10^{-3} moles of deformable molecules per unit volume. Use the average of the cosines of the angles between the dipole moment of the central molecule and that of its bonded neighbors as 1/3. (Cf. Exercise 2.32 in the textbook) (Contractor-GamboaAldeco)

Answer:

(a) Gauss's law says that the electric field normal to the surface (Gaussian surface) of any volume is $1/\varepsilon_0$ times the charge in the volume. The volume chosen for this problem is a brick-shaped volume with two of its faces of unit area parallel to the capacitor plate, and enclosing the charges q_M and $-q_{dipole}$ (see Fig. 2.12). The net charge in the volume is $q_M - q_{dipole}$ and the net electric field is

$$\vec{X}_{ext} = \frac{q_M - q_{dipole}}{\varepsilon_0} \tag{2.224}$$

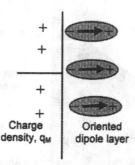

Figure 2.11. Dielectric molecules at a metallic plate of a capacitor.

(b) From Eq. (2.219), $q_{dipole} = 4\pi\varepsilon_0 n\alpha \vec{X}_{ext}$, then

$$\vec{X}_{ext} = \frac{q_M - 4\pi\varepsilon_0 n\alpha \vec{X}_{ext}}{\varepsilon_0} \tag{2.225}$$

or rearranging terms,

$$\vec{X}_{ext} = \frac{q_M}{\varepsilon_0 + 4\pi\varepsilon_0 n\alpha} \tag{2.226}$$

Substituting $\vec{X}_{ext} = V/d$ and rearranging Eq. (2.226),

$$\frac{q_M}{V} = \frac{\varepsilon_0 + 4\pi\varepsilon_0 n\alpha}{d} \tag{2.227}$$

For a parallel-plate condenser $C = q/V = \varepsilon\varepsilon_0/d$, then,

$$\frac{q_M}{V} = \frac{\varepsilon_0 + 4\pi\varepsilon_0 n\alpha}{d} = \frac{\varepsilon\varepsilon_0}{d} \tag{2.228}$$

or

$$\varepsilon = 1 + 4\pi n\alpha \tag{2.229}$$

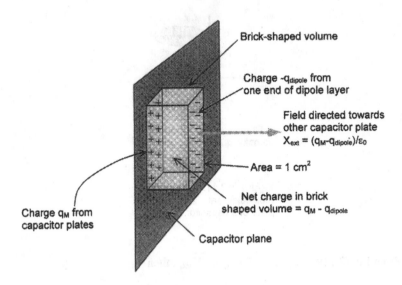

Figure 2.12. Gaussian surface.

(c) With $q_{dipole} = q_{perm} + q_{ind}$, then

$$\alpha = \frac{q_{dipole}}{4\pi\varepsilon_0 n \vec{X}_{ext}} = \frac{q_{perm}}{4\pi\varepsilon_0 n \vec{X}_{ext}} + \frac{q_{ind}}{4\pi\varepsilon_0 n \vec{X}_{ext}} \qquad (2.230)$$

where

$$q_{perm} = n \left\langle \mu_{group} \right\rangle = n \frac{\mu^2 \left(1 + g\,\overline{cos\gamma}\right)^2}{3kT} \frac{3\varepsilon}{2\varepsilon+1} \vec{X}_{ext} \qquad (2.231)$$

and

$$q_{ind} = 4\pi\varepsilon_0 n \alpha_{ind} \frac{3\varepsilon}{2\varepsilon+1} \vec{X}_{ext} \qquad (2.232)$$

Therefore,

$$\alpha = \frac{\mu^2 \left(1 + g\,\overline{cos\gamma}\right)^2}{4\pi\varepsilon_0 \; 3kT} \frac{3\varepsilon}{2\varepsilon+1} + \alpha_{ind} \frac{3\varepsilon}{2\varepsilon+1} \qquad (2.233)$$

Substituting Eq. (2.233) in Eq. (2.229),

$$\varepsilon = 1 + \frac{n\mu^2 \left(1 + g\overline{\cos\gamma}\right)^2}{\varepsilon_0 \, 3kT} \frac{3\varepsilon}{2\varepsilon + 1} + \alpha_{ind} \frac{4\pi n 3\varepsilon}{2\varepsilon + 1} = 1 + \frac{n}{\varepsilon_0} \frac{3\varepsilon}{2\varepsilon + 1}$$

$$\times \left(\frac{\mu^2 \left(1 + g\overline{\cos\gamma}\right)^2}{3kT} + 4\pi\varepsilon_0 \alpha_{ind} \right) \qquad (2.234)$$

or

$$\frac{(\varepsilon - 1)(2\varepsilon + 1)}{3\varepsilon} = \frac{n}{\varepsilon_0} \left[\frac{\mu^2 \left(1 + g\overline{\cos\gamma}\right)^2}{3kT} + 4\pi\varepsilon_0 \alpha_{ind} \right] \qquad (2.235)$$

This is Kirkwood's equation for the dielectric constant of a condensed medium. It takes into account the short range interactions between polar molecules that lead to the formation of molecular groups oriented as a unit under the influence of electric fields.

(d) Using Kirkwood 's equation,

$$\alpha_{ind} = \frac{(\varepsilon - 1)(2\varepsilon + 1)}{12\pi\varepsilon n} - \frac{\mu^2 \left(1 + g\overline{\cos\gamma}\right)^2}{12\pi\varepsilon_0 kT}$$

$$= \frac{(78.5 - 1)(157 + 1)}{12\pi(78.5)\left(10^{-3} \, mol \, cm^{-3}\right)6.022 \times 10^{23}} \frac{1 \, mol}{} \qquad (2.236)$$

$$- \left\{ \frac{(1.87D)^2 \left(1 + 4\left(\frac{1}{3}\right)\right)^2}{3\left(1.112 \times 10^{-10} \, C^2 J^{-1} m^{-1}\right)\left(1.381 \times 10^{-23} \, JK^{-1} \, 298 \, K\right)} \right.$$

$$\left. \times \frac{\left(3.336 \times 10^{-30} \, Cm\right)^2}{D^2} \frac{(100 \, cm)^3}{(1m)^3} \right\} = 6.71 \times 110^{-21} \, cm^3$$

MICRO-RESEARCH PROBLEMS

2.56 The origin of the hydrogen bond is the intermolecular dipole interaction caused by the polarized covalent bond. The existence of this additional intermolecular force accounts for the abnormally high boiling point of water (cf. Section 2.4 in the textbook). **(a)** Using the electronegativity (χ) data provided below, as well as the boiling points (b.p.) of different hydrides in Figure 2.13, rationalize semi-quantitatively the elevation of boiling point of NH_3, H_2O, and HF in terms of bond polarization. Consider the total intermolecular force caused by H-bond to be proportional to the total number of H bonds times the dipole moment of an individual H bond, which in turn varies approximately with the difference in electronegativity between the hydrogen and the anion, i.e., $\Delta\chi$ $= \chi_X - \chi_{H}$.

	H	N	O	F	Zn
Electronegativity, χ	2.2	3.0	3.4	4.0	1.6

(b) The elevation of boiling point is by no means unique to hydrogen. Explain the abnormal boiling point of ZnF_2 (Fig. 2.14) in the light of the "zinc bond". Compare the "zinc bond" with the hydrogen bond in the hydrides studied above and account for the especially strong effect of the "zinc bond" on the boiling point. Assume that each ZnF_2 molecule can form six "zinc bonds" (this number is derived from the largest possible coordination number zinc could have). (Cf. Micro research problem 2.2 in the textbook) (Xu)

Answer:

(a) To a first approximation, we can suppose that the boiling point elevation is proportional to the total effect of H-bond on the boiling point. This effect can be accounted as the product of the *polarization* of single H-X bond times the *number* of H-bonds formed. The polarization is approximately given by $\Delta\chi$, while the number of hydrogen bonds is known from structural studies. Now, the *increase* in boiling point for every hydrogen bond of unit dipole moment should be independent of the hydride, that is

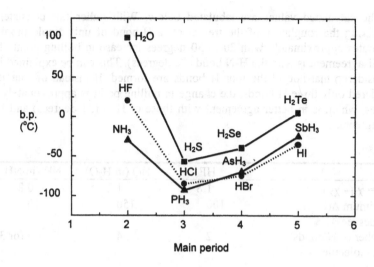

Figure 2.13. Boiling points of hydrides

$$\Delta(\mathrm{b.p.})\left[\mathrm{degrees}(\mathrm{H}-\mathrm{bond})^{-1}(\Delta\chi)^{-1}\right]$$

$$=\frac{\Delta(\mathrm{b.p.})}{(\mathrm{numberof\ H}-\mathrm{bonds})(\Delta\chi)}=\mathrm{constant} \qquad (2.237)$$

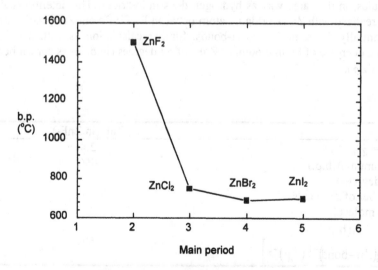

Figure 2.14. Boiling points of zinc halides.

The calculated values are tabulated below. With rather fair consistency considering the roughness of the treatment, a H-bond of unit dipole moment contributes approximately with 20 to 30 degrees increase in boiling point. The only disagreement is with the H-N bond (22 degrees). This can be explained if it is considered that not all the four H bonds are formed. If instead of four it is considered only three H bonds, the change in boiling point is approximately 29 degrees, which is in better agreement with those of H-O (31 degrees) and H-F (28 degrees).

	HF	HO (in H_2O)	NH (in NH_3)
$\Delta\chi$ $(= \chi_H - \chi_x)$	1.8	1.2	0.8
Maximum Δ(b.p.) (degrees)	100	150	70
Number of H bonds in molecule	2	4	4 (or 3)
$\dfrac{\Delta(\text{b.p.})}{\left[\deg(\text{H}-\text{bond})^{-1}(\Delta\chi)^{-1}\right]}$	27.8	31.2	21.8 (or 29.1)

(b) Because the Zn-F is severely polarized due to the big difference between the atom electronegativities, Zn is rendered as very "electron thirsty". As a result, Zn begins to interact with neighbor electron-rich fluorides of neighboring ZnF_2 molecules, in the same way as hydrogen does in hydrides. This intermolecular force, realized with Zn as bridging atom between F of different ZnF_2 molecules, is essentially the same as the H-bonds. Similar calculation as with hydrides shows a decrease of boiling point of ZnF_2 of 52 degrees (in degrees per Zn bond and per $\Delta\chi$):

	ZnF (in ZnF_2)
$\Delta\chi$ $(= \chi_{Zn} - \chi_x)$	2.4
Maximum Δ(b.p.) (degrees)	650
Number of Zn bonds in molecule	6(?)
$\dfrac{\Delta(\text{b.p.})}{\left[\deg(\text{Zn}-\text{bond})^{-1}(\Delta\chi)^{-1}\right]}$	52

The value of $\Delta(b.p.)$ for ZnF_2 is much higher than any of the above hydrides (compare both tables). The reason might be the partial ionic nature of the Zn-F bond. The difference in electronegativity is so large, that this may indicate that the Zn-F bond is partially dissociated, that is, it is partially electrostatic in nature. This means that the Zn bond is to some extent non-directional, and works in a much wider range, in contrast to the H bond, which is directional and stoichiometrical, that is, only affects the nearest neighbors. As a consequence, much more F atoms interact with a given Zn atom (in comparison with a given H atom), explaining the further increase in boiling point.

CHAPTER 3

ION-ION INTERACTIONS

EXERCISES

Review of Sections 3.1 to 3.3 of the Textbook.

Describe and give examples of *ionophore, ionogen, strong* and *waek electrolyte*. Why is the charge density around a reference ion not *zero* when the whole solution is electroneutral? Explain succinctly the Debye-Hückel approach to their ionic-cloud theory. Mention the main steps involved in this theory as well as the assumptions involved in its development. Write *the linearized Poisson-Boltzmann equation* and the time-average spatial distribution of the excess charge density around a reference ion. What is the total ionic-cloud charge around a reference ion? Draw the variation of the electrostatic potential and the excess-charge density as a function of the distance from the central ion. What does κ^{-1} represent? Write expressions for the Debye-Hückel length, the total electrostatic potential, the potential due to the isolated central ion, and the potential due to the ionic cloud. What is the physical meaning of *effective thickness of the ion atmosphere*?

3.1 (a) Write down the full Taylor expansion of e^x and e^{-x}. (b) Find the percentage difference between e^x and $1+x$ for $x = 0.1$, 0.5 and 0.9. (Cf. Exercise 3.32 in the textbook) (Bockris-GamboaAldeco)

Answer:

(a) The expansions are:

$$e^x = 1 + x + \frac{x^2}{2!} + \frac{x^3}{3!} + \dots \qquad (3.1)$$

$$e^{-x} = 1 - x + \frac{x^2}{2!} - \frac{x^3}{3!} + \dots \qquad (3.2)$$

(b)

x	e^x	$1+x$	$\dfrac{e^x - (1+x)}{e^x} \times 100$
0.1	1.1052	1.1	0.47 %
0.5	1.6487	1.5	9.02 %
0.9	2.4596	1.9	22.57 %

3.2 Evaluate the electrostatic potential ψ and the work of charging Cu^{+2} ions from a state of zero charge to a charge of $z_i e_0$ in water at 25 ^{0}C. (Contractor)

Data:

$z_i = 2$ $T = 298$ K

$\varepsilon = 78.3$ $r_{Cu^{+2}} = 72$ pm

Answer:

The work of charging an electrical conductor is (cf. Eq. 3.3 in the textbook),

$$W = \frac{z_i e_0}{2} \psi \qquad (3.3)$$

and, the electrostatic potential ψ, is given by the equation (cf. Eq. 3.30 in the textbook),

$$\psi = \frac{z_i e_0}{4 \pi \varepsilon_0 \varepsilon r_i} \qquad (3.4)$$

$$= \frac{(2)\left(1.602\times10^{-19}\ \text{C}\right)}{\left(1.112\times10^{-10}\ \text{C}^2\text{J}^{-1}\text{m}^{-1}\right)(78.3)\left(72\times10^{-12}\ \text{m}\right)} \times \frac{1\ \text{VC}}{1\ \text{J}} = 0.51\ \text{V}$$

Substituting this value into Eq. (3.3),

$$W = \frac{(2)\left(1.602\times10^{-19}\ \text{C}\right)}{2}(0.51\ \text{V})\times\frac{1\ \text{J}}{1\ \text{VC}} = 8.18\text{x}10^{-20}\ \text{J} \quad (3.5)$$

3.3 (a) The excess charge of an ionic atmosphere varies with distance out from the central ion. Starting from the expression for the excess charge density, show that the net change in a spherical shell of thickness dr is

$$dq = -z_i e_0 e^{-\kappa r} \kappa^2 r dr \quad (3.6)$$

(b) At a certain distance from a central ion, there will be a ring with a maximum charge. Find the distance of this ring from the central ion. (Cf. Problem 3.7 in the textbook) (Bockris-GamboaAldeco)

Answer:

(a) The excess charge density is given by (cf. Eq. 3.35 in the textbook),

$$\rho_r = -\frac{z_i e_0 \kappa^2}{4\pi} \frac{e^{-\kappa r}}{r} \quad (3.7)$$

Therfore, the total charge in a spherical shell of thickness dr at a distance r from the reference ion is given by ρ_r times the volume in this shell, i.e., $4\pi r^2 dr$, or,

$$dq = 4\pi\rho_r r^2 dr \quad (3.8)$$

Substituting ρ_r from Eq. (3.7) into Eq. (3.8) gives the equation we are looking for, i.e.,

$$dq = -z_i e_0 \kappa^2 r e^{-\kappa r} dr \quad (3.9)$$

(b) The excess charge in a spherical shell of thickness dr, that is dq/dr, has a maximum when $\dfrac{d}{dr}\left(\dfrac{dq}{dr}\right)=0$. Thus, differentiating Eq. (3.9),

$$\frac{d^2q}{dr^2}=\frac{d}{dr}\frac{dq}{dr}=\frac{d}{dr}\left(-z_i e_0 \kappa^2 re^{-\kappa r}\right) \tag{3.10}$$

$$=-z_i e_0 \kappa^2 \left(e^{-\kappa r}-r\kappa e^{-\kappa r}\right)=0$$

$$e^{-\kappa r}=r\kappa e^{-\kappa r} \tag{3.11}$$

or

$$r=\kappa^{-1} \tag{3.11a}$$

3.4 The potential of the ionic atmosphere according to the limiting law (i.e., low concentration) is given by $\psi_{cloud}=-z_i e_0 \kappa / \varepsilon$, **where** κ^{-1} **is the so-called Debye-Hückel length. Explain why** κ^{-1} **is called in this way and draw a figure showing the meaning of this term in terms of the concept of "ionic atmosphere." (Cf. Exercise 3.27 in the textbook) (Bockris-GamboaAldeco)**

Answer:

The term κ^{-1} represents the distance at which a spherical shell around an ion contains the maximum value of charge. Since κ^{-1} represents a distance and has units of length, it is referred to as the *Debye-Hückel length*. A scheme showing this concept is given in Fig. 3.1. In this figure, dq represents the charge enclosed in a dr-thick spherical shell, and r the distance from the reference ion.

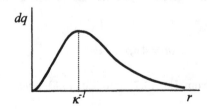

Figure 3.1. Charge in a dr-thick spherical shell as a function of distance.

$$\kappa^{-1} = \left(\frac{\varepsilon \varepsilon_0 kT}{1000 e_0^2 N_A \sum_i c_i z_i^2} \right)^{1/2} \tag{3.19}$$

For a 1:1, 0.1 M electrolyte, $\sum_i c_i z_i^2 = c(1)^2 + c(-1)^2 = 2c = 0.2 \, \text{M}$.

Therefore,

$$\kappa^{-1} = \left[\frac{\varepsilon \left(8.854 \times 10^{-12} \, \text{C}^2 \text{J}^{-1} \text{m}^{-1} \right) \left(1.381 \times 10^{-23} \, \text{JK}^{-1} \right)}{\left(1000 \, \text{dm}^3 \text{m}^{-3} \right) \left(1.602 \times 10^{-19} \, \text{C} \right)^2 \left(6.023 \times 10^{23} \, \text{mol}^{-1} \right)} \right]^{1/2} \tag{3.20}$$

$$\times \frac{298 \text{K}}{0.2 \, \text{mol dm}^{-3}} \bigg]^{1/2} = \sqrt{\varepsilon} \, 1.086 \times 10^{-10} \, \text{m}$$

Substituting the different values of ε for the different solvents gives

Solvent	ε	κ^{-1} (m)
Nitrobenzene	34.8	6.404×10^{-10}
Ethyl alcohol	24.3	5.352×10^{-10}
Ethylen dichloride	10.4	3.501×10^{-10}

3.7 Calculate the potentials due to the ionic cloud around cations in the following aqueous solutions at 25 °C: 10^{-3} M NaCl, 0.1 M NaCl, 10^{-3} M CaCl$_2$, and 10^{-3} M CaSO$_4$. Considering your results from this problem as well as those of Exercise 3.6, what conclusions can you make with respect to the variation of the dielectric constant and the potential of the ionic cloud? (Cf. Exercise 3.1 in the textbook) (Kim)

Answer:

The potential due to the ionic cloud is obtained from the following equation written in the *mksa* system (cf. Eq. 3.49 in the textbook),

$$\psi_{cloud} = -\frac{z_i e_0}{4\pi\varepsilon_0 \varepsilon \kappa^{-1}} \tag{3.21}$$

where κ^{-1} is (cf. Eq. 3.14 in Exercise 3.5):

$$\kappa^{-1} = \left(\frac{\varepsilon\varepsilon_0 kT}{e_0^2 \sum_i n_i^0 z_i^2}\right)^{1/2} \tag{3.22}$$

For 10^{-3} M NaCl,

$$\sum_i n_i^0 z_i^2 = n_{Na^+}^0 z_{Na^+}^2 + n_{Cl^-}^0 z_{Cl^-}^2$$

$$= \left[\left(10^{-3} \text{ mol dm}^{-3}\right)(+1)^2 + \left(10^{-3} \text{ mol dm}^{-3}\right)(-1)^2\right] \tag{3.23}$$

$$\times \left(6.022 \times 10^{23} \text{ mol}^{-1}\right)\frac{1000 \text{ dm}^3}{1 \text{ m}^3} = 1.205 \times 10^{24} \text{ m}^{-3}$$

Substituting the corresponding values into Eq. (3.22),

$$\kappa^{-1} = \left[\frac{(78.3)\left(8.854 \times 10^{-12} \text{ C}^2 \text{ J}^{-1} \text{ m}^{-1}\right)}{\left(1.602 \times 10^{-19} \text{ C}\right)^2}\right]$$

$$\left.\frac{\left(1.381 \times 10^{-23} \text{ JK}^{-1}\right)(298 \text{ K})}{\left(1.205 \times 10^{24} \text{ m}^{-3}\right)}\right]^{1/2} = 9.607 \times 10^{-9} \text{ m} \tag{3.24}$$

Substituting now Eq. (3.24) into Eq. (3.21),

$$\psi_{cloud} = -\frac{(+1)\left(1.602 \times 10^{-19} \text{ C}\right)}{\left(1.112 \times 10^{-10} \text{ C}^2 \text{ J}^{-1} \text{ m}^{-1}\right)(78.3)\left(9.607 \times 10^{-9} \text{ m}\right)} \tag{3.25}$$

$$\times \frac{1 \text{V C}}{1 \text{J}} = 1.915 \text{x} 10^{-3} \text{ V}$$

In the same way for the other solutions,

	$\sum_i n_i^0 z_i^2$	κ^{-1}	ψ_{cloud}
	(m^{-3})	(m)	(V)
10^{-3} M NaCl	$1.205 \text{x} 10^{24}$	$9.607 \text{x} 10^{-9}$	$1.915 \text{x} 10^{-3}$
10^{-1} M NaCl	$1.205 \text{x} 10^{26}$	$9.607 \text{x} 10^{-10}$	$1.915 \text{x} 10^{-2}$
10^{-3} M $CaCl_2$	$3.614 \text{x} 10^{24}$	$5.546 \text{x} 10^{-9}$	$6.635 \text{x} 10^{-3}$
10^{-3} M $CaSO_4$	$4.818 \text{x} 10^{24}$	$4.803 \text{x} 10^{-9}$	$7.661 \text{x} 10^{-3}$

Comment: The thickness of the ionic atmosphere increases with increasing the dielectric constant, ε, of the solvent. It decreases with increasing the electrolyte concentration and with increasing the valence of the ions in the electrolyte. On the other hand, ψ_{cloud} increases with the increase of the concentration of the electrolyte and with the valence of the ions.

3.8 Consider a 0.001 N aqueous KCl solution. Calculate the potentials at distances $\kappa^{-1}/2$, κ^{-1} and 3 κ^{-1} due to the K^+ ion, its ionic atmosphere, and the total potential. (Cf. Exercise 3.2 in the textbook) (Bockris-GamboaAldeco)

Answer:

The value of κ^{-1} for a 1:1, 10^{-3} M electrolyte was calculated in Exercise 3.7 as $9.607 \text{x} 10^{-9}$ m. The potential at a distance $r = \kappa^{-1}/2$ due to the K^+ ion, is given in the *mksa* system by (cf. Eq. 3.44 in the textbook):

$$\psi_{ion} = \frac{z_i e_0}{4 \pi \varepsilon_0 \varepsilon r} = \frac{2(+1)\left(1.602 \times 10^{-19} \text{ C}\right)}{\left(1.112 \times 10^{-10} \text{ C}^2 \text{J}^{-1}\text{m}^{-1}\right)(78.3)\left(9.607 \times 10^{-9} \text{ m}\right)}$$

$$\times \frac{1 \text{VC}}{1 \text{J}} = 0.00383 \text{V} \tag{3.26}$$

The potential due to the ionic cloud is given in the *mksa* system by (cf. Eq. 3.49 in the textbook):

$$\psi_{cloud} = -\frac{z_i e_0}{4\pi\varepsilon_0 \varepsilon\kappa^{-1}}$$

$$= \frac{(+1)\left(1.602\times10^{-19}\ C\right)}{\left(1.112\times10^{-10}\ C^2 J^{-1} m^{-1}\right)(78.3)\left(9.607\times10^{-9}\ m\right)}\times\frac{1VC}{1J} \quad (3.27)$$

$$= -0.00192\ V$$

The total potential at a distance r from the ion is given by the sum of the potentials due to the central ion and its ionic cloud, i.e., (cf. Eq. 3.45) in the textbook),

$$\psi_{total} = 0.00383\ V - 0.00192\ V = +0.00192\ V \quad (3.28)$$

Using the same equations, the values of ψ_{ion}, ψ_{cloud}, and ψ_{total} for $r = \kappa^{-1}$ and $3\kappa^{-1}$ are,

r	ψ_{ion} (V)	ψ_{cloud} (V)	ψ_{total} (V)
$0.5\kappa^{-1}$	+0.00383	-0.00192	+0.00192
κ^{-1}	+0.00192	-0.00192	0
$3\ \kappa^{-1}$	+0.00064	-0.00192	-0.00128

Comment: At κ^{-1} the effect of the ion cloud is equivalent to the effect of the central ion, with opposite charge, and both effects cancel out each other.

3.9 Calculate the change of chemical potential of the following solutions: 10^{-3} M NaCl, 0.1 M NaCl, 10^{-3} M CaCl$_2$, and 10^{-3} M CaSO$_4$. Consider the Debye-Hückel limiting law at 25 °C applies. Use the results from Exercise 3.7. (Cf. Exercise 3.18 in the textbook) (Kim)

Data:

	ψ_{cloud} (V)		ψ_{cloud} (V)
10^{-3} M NaCl	1.915×10^{-3}	10^{-3} M CaCl$_2$	6.635×10^{-3}
10^{-1} M NaCl	1.915×10^{-2}	10^{-3} M CaSO$_4$	7.661×10^{-3}

Answer:

The chemical potential difference is given in the *mksa* unit system by (cf. Eqs. 3.3 and 3.49 in the textbook):

$$\Delta\mu_{i-I} = \frac{N_A z_i e_0}{2}\psi_{cloud} \qquad (3.29)$$

Therefore, for Na^+ in 10^{-3} M NaCl,

$$\Delta\mu_{i-I} = \frac{\left(6.022\times10^{23}\ mol^{-1}\right)(+1)\left(1.602\times10^{-19}\ C\right)}{2}$$
$$\times\left(1.915\times10^{-3}\ V\right)\times\frac{1J}{1CV} = -92.39\ \text{J mol}^{-1} \qquad (3.30)$$

Similarly,

	c (M)	ψ_{cloud} (V)	$\Delta\mu_{i-I}$ (J mol^{-1})
Na^+ (NaCl)	0.001	1.915×10^{-3}	- 92.39
Na^+ (NaCl)	0.1	1.915×10^{-2}	-923.9
Ca^{+2} (CaCl$_2$)	0.001	6.635×10^{-3}	-640.2
Ca^{+2} (CaSO$_2$)	0.001	7.661×10^{-3}	-739.2

3.10 Derive the linearized Poisson-Boltzmann equation. (Cf. Exercise 3.33 in the textbook) (Bockris)

Answer:

The excess charge density in the volume element dV is given by (cf. Eq. 3.10 in the textbook):

$$\rho_r = \sum_i n_i^0 z_i e_0 e^{-z_i e_0 \psi_r / kT} \qquad (3.31)$$

where n_i^0 is the number of ions in the bulk, z_i the ion charge, ψ_r the electrostatic potential in the volume element dV. Thus, when

$$z_i e_0 \psi_r \ll kT \quad \text{(Debye-Hückel assumption)} \qquad (3.32)$$

the exponential in Eq. (3.31) can be expanded in a Taylor series. If only the first term of the series is taken into account, then,

$$e^{-z_i e_0 \psi_r / kT} \approx 1 - \frac{z_i e_0 \psi_r}{kT} \tag{3.33}$$

Substituting Eq. (3.33) into Eq. (3.31),

$$\rho_r = \sum_i n_i^0 z_i e_0 \left(1 - \frac{z_i e_0 \psi_r}{kT} \right) = \sum_i n_i^0 z_i e_0 - \sum_i \frac{n_i^0 z_i^2 e_0^2 \psi_r}{kT}$$

$$= - \sum_i \frac{n_i^0 z_i^2 e_0^2 \psi_r}{kT} \tag{3.34}$$

where the first term, $\sum_i n_i^0 z_i e_0$, has been removed because it represents the

charge in the electrolyte solution as a whole and is equal to zero. Equation (3.33) is the *linearized-Boltzmann equation*. On the other hand, the Poisson equation for a spherical symmetrical charge distribution is (cf. Eq. 3.4 in the textbook),

$$\rho_r = -\frac{\varepsilon}{4\pi} \frac{1}{r^2} \frac{d}{dr} \left(r^2 \frac{d\psi_r}{dr} \right) \tag{3.35}$$

Equating Eq. (3.34) and (3.35),

$$\frac{\varepsilon}{4\pi} \frac{1}{r^2} \frac{d}{dr} \left(r^2 \frac{d\psi_r}{dr} \right) = \sum_i \frac{n_i^0 z_i^2 e_0^2 \psi_r}{kT} \tag{3.36}$$

Defining $\kappa^2 \equiv \frac{4\pi}{\varepsilon kT} \sum_i n_i^0 z_i^2 e_0^2$, and substituting it into Eq. (3.36),

$$\frac{1}{r^2} \frac{d}{dr} \left(r^2 \frac{d\psi_r}{dr} \right) = \kappa^2 \psi_r \tag{3.37}$$

and this is the linearized Poisson-Boltzmann equation.

Review of Section 3.4 in the Textbook.

Define *activity coefficient* and mention its relationships to the *activity* and the *chemical potential* of species i. Define the *mean ionic-activity coefficient*. What is *the ionic strength* of a solution? Mention two methods to determine solute activities and their drawbacks.

3.11 (a) Compare the ionic strengths of 1:1, 2:1, 2:2 and 3:1 valent electrolytes in solutions of molarity c. (b) Based on these results, find a relationship between the ionic strength and molarity. (Cf. Exercise 3.5 in the textbook) (Constantinescu)

Answer:

(a) The ionic strength for a single electrolyte solution is given by (cf. Eq. 3.83 in the textbook),

$$I = \frac{1}{2}\sum_i c_i z_i^2 = \frac{1}{2}\left(c_+ z_+^2 + c_- z_-^2 \right) \qquad (3.38)$$

Applying Eq. (3.38) to the given solutions :

Type	z_+	z_-	c_+	c_-	I
1:1 (AB)	1	1	c	c	$\frac{1}{2}\left[c(1)^2 + c(1)^2 \right] = c$
2:1 (AB_2)	2	1	c	$2c$	$\frac{1}{2}\left[c(2)^2 + 2c(1)^2 \right] = 3c$
2:2 (AB)	2	2	c	c	$\frac{1}{2}\left[c(2)^2 + c(2)^2 \right] = 4c$
3:1 (AB_3)	3	1	c	$3c$	$\frac{1}{2}\left[c(3)^2 + 3c(1)^2 \right] = 6c$

(b) From the above results,

$$\mathbf{I = constant \times concentration} \qquad (3.39)$$

3.12 Calculate the ionic strength of the following solutions: (a) 0.04 M KBr, (b) 0.35 M BaCl$_2$ and (c) 0.02 M Na$_2$SO$_4$ + 0.004 M Na$_3$PO$_4$ + 0.01 M AlCl$_3$. (Cf. Exercise 3.6 in the textbook) (Constantinescu)

Answer:

From the results of Exercise 3.11,

(a) 0.04 M KBr is a 1:1 (AB) type electrolyte and thus, $I = c = 0.04$ M

(b) 0.35 M BaCl$_2$ is a 2:1 (AB$_2$) type electrolyte and thus, $I = 3c = 3(0.35) = 1.05$ M

(c) 0.02 M Na$_2$SO$_4$ + 0.004 M Na$_3$PO$_4$ + 0.01 M AlCl$_3$ is a mixture of electrolytes, thus (cf. Eq. 3.83 in the textbook),

$$I = \frac{1}{2}\sum_i c_i z_i^2 \tag{3.40}$$

$$= \frac{1}{2}\left[c_{Na^+}(1)^2 + c_{SO_4^{-2}}(2)^2 + c_{PO_4^{-3}}(3)^2 + c_{Al^{+3}}(3)^2 + c_{Cl^-}(1)^2 \right]$$

$$= \frac{1}{2}\left[0.052(1)^2 + 0.02(4) + 0.004(9) + 0.01(9) + 0.03(1)^2 \right] = 0.144 \text{M}$$

3.13 Calculate the mean ionic-activity coefficients of the following salts in aqueous solutions at 298 K: 10^{-3} M NaCl, 0.1 M NaCl, 10^{-3} M CaCl$_2$, and 10^{-3} M CaSO$_4$. Consider the Debye-Hückel limiting law at 25 ^{0}C. (Cf. Exercise 3.9 in the textbook) (Kim)

Answer:

The activity coefficient according to the Debye-Hückel limiting law (cf. Eq. 3.88 in the textbook) is given by the following expression in the *mksa* unit system,

$$log\, f_{\pm} = -\frac{1}{2.303(4\pi\varepsilon_0)}\frac{e_0^2}{2\varepsilon kT} B(z_+ z_-) I^{1/2} \tag{3.41}$$

The parameter B in the *mksa* unit system is (cf. Eq. 3.86 in the textbook),

$$B = \left(\frac{2N_A e_0^2}{\varepsilon \varepsilon_0 kT}\right)^{1/2} = \left[\frac{2\left(6.022 \times 10^{23} \, \text{mol}^{-1}\right)}{(78.3)\left(8.854 \times 10^{-12} \, \text{C}^2 \text{J}^{-1} \text{m}^{-1}\right)}\right.$$

$$\left.\times \frac{\left(1.602 \times 10^{-19} \, \text{C}\right)^2}{\left(1.381 \times 10^{-23} \, \text{JK}^{-1}\right)(298\,\text{K})}\right]^{1/2} = 1.041 \times 10^8 \, \text{mol}^{-1/2} \, \text{m}^{1/2}$$

(3.42)

or,

$$B = 1.041 \times 10^8 \, \text{mol}^{-1/2} \text{m}^{1/2} \frac{\text{dm}^{3/2}}{\text{dm}^{3/2}} \left(\frac{1000\,\text{dm}^3}{1\,\text{m}^3}\right)^{1/2}$$

(3.43)

$$= 3.291 \times 10^9 \, \text{M}^{-1/2} \text{m}^{-1}$$

Substituting this value of B in Eq. (3.41)

$$\log f_\pm = -\frac{\left(1.602 \times 10^{-19} \, \text{C}\right)^2}{2.303\left(1.112 \times 10^{-10} \, \text{C}^2 \text{J}^{-1} \text{m}^{-1}\right)}$$

$$\times \frac{\left(3.291 \times 10^9 \, \text{M}^{-1/2} \text{m}^{-1}\right)(z_+ z_-)I^{1/2}}{2(78.3)\left(1.38 \times 10^{-23} \, \text{JK}^{-1}\right)(298\,\text{K})}$$

(3.44)

$$= \left(0.512\,\text{M}^{-1/2}\right)(z_+ z_-)I^{1/2}$$

Therefore, for 10^{-3} M NaCl, $z_+ = 1$, $z_- = -1$ and $I = c = 10^{-3}$ M (cf. Exercise 3.11)

$$\log f_\pm = -\left(0.512\,\text{M}^{-1/2}\right)|1||-1|\left(10^{-3}\right)^{1/2} = -0.0162$$

$$\Rightarrow \quad f_\pm = 0.963$$

(3.45)

Similarly for the other electrolytes,

Solution	I		$\log f_\pm$	$f_\pm$
10^{-3} M NaCl (1:1)	$I = c =$	10^{-3} M	-0.0162	0.963
10^{-1} M NaCl (1:1)	$I = c =$	10^{-1} M	-0.162	0.689
10^{-3} M CaCl$_2$ (2:1)	$I = 3c = 3 \times 10^{-3}$ M		-0.0561	0.879
10^{-3} M CaSO$_4$ (2:2)	$I = 4c = 4 \times 10^{-3}$ M		-0.129	0.743

3.14 What are the units of the constants A and B used in the Debye-Hückel theory when the equations are given in the *mksa* system, i.e.

$$\log f_\pm = -A(z_+ z_-)I^{1/2} \tag{3.46}$$

where

$$A = \frac{1}{2.303} \frac{N_A e_0^2}{8 \pi \varepsilon_0 \varepsilon RT} B \tag{3.47}$$

and

$$B = \left(\frac{2 N_A e_0^2}{\varepsilon \varepsilon_0 kT} \right)^{1/2} \tag{3.48}$$

Are these the units given in Tables 3.3 and 3.4 in the textbook? (GamboaAldeco)

Answer:

From Eqs. (3.42) and (3.43) in Exercise 3.13, the units of B in the *mksa* system are

$$[B] \rightarrow \text{mol}^{-1/2}\ \text{m}^{1/2} \quad \text{or} \quad \text{M}^{-1/2}\ \text{m}^{-1} \tag{3.49}$$

However, the units in Table 3.3 in the textbook are expressed in $[B] \rightarrow \text{M}^{-1/2}\ \text{cm}^{-1}$. For example, from Exercise 3.13, at 25 ^{0}C, B = 3.291×10^9 M$^{-1/2}$ m^{-1}. Converting m^{-1} into cm^{-1} gives,

$$B = 3.291 \times 10^9 \ M^{-1/2} m^{-1} \times \frac{1m}{100 \, cm} = 0.3291 \times 10^8 \ M^{-1/2} cm^{-1} \quad (3.50)$$

which is the value given in the textbook for B at this temperature. In the *mksa* system, A is

$$[A] \rightarrow \left[\frac{mol^{-1} C^2}{\left(C^2 J^{-1} m^{-1} \right) \left(JK^{-1} mol^{-1} \right) K} \left(mol^{-1/2} m^{1/2} \right) \right] \quad (3.51)$$

$$\rightarrow mol^{-1/2} m^{3/2}$$

The units of A in Table 3.4 in the textbook are $A \rightarrow M^{-1/2}$. In our same example at 25 ^{0}C,

$$[A] = \frac{1}{2.303} \frac{\left(6.023 \times 10^{23} \ mol^{-1} \right) \left(1.602 \times 10^{-19} \ C \right)^2}{2 \left(1.112 \times 10^{-10} \ C^2 J^{-1} m^{-1} \right) (78.3) \left(8.314 \, J \, mol^{-1} \right) (298 \, K)}$$

$$\times \left(1.041 \times 10^8 \ mol^{-1/2} m^{1/2} \right)$$

$$= 0.01619 \, mol^{-1/2} m^{3/2} \times \frac{dm^{3/2}}{dm^{3/2}} \frac{\left(1000 \, dm^3 \right)^{1/2}}{1 m^{3/2}} = 0.512 \, M^{-1/2} \quad (3.52)$$

This is the value given for A in Table 3.4 in the textbook at 25 ^{0}C.

3.15 (a) Evaluate the Debye-Hückel reciprocal length, κ^{-1} in 0.001 M solution of NaCl in water at 25 ^{0}C. Use the constant $B = 0.3291 \times 10^8 \ M^{-1/2}$ cm^{-1}. **(b)** How does κ^{-1} varies if the concentration of this solution is doubled keeping all the other conditions the same? **(c)** How does κ^{-1} varies if instead of NaCl the solution under study is 0.001 M CdCl$_2$? (Contractor)

Answer:

(a) The Debye-Hückel reciprocal length is given in terms of the constant B (cf. Eq. 3.85 in the textbook),

$$\kappa = B\sqrt{I} \tag{3.53}$$

For a 1:1 electrolyte, $I = c$ (cf. Exercise 3.11), and from the textbook the value of B at 298 K is 0.3291×10^8 M$^{-1/2}$ cm^{-1}. Therefore,

$$\kappa = 0.3291 \times 10^8 \text{ M}^{-1/2} \text{cm}^{-1} \sqrt{0.001 \text{M}} = 1.041 \times 10^6 \text{ cm}^{-1} \tag{3.54}$$

or

$$\kappa^{-1} = 9.609 \times 10^{-7} \text{ cm} = \textbf{9.609x10}^{-9} \textbf{ m} \tag{3.55}$$

(b) Doubling the concentration of NaCl will result in doubling the ionic strength ($I = 2 c_1$). Therefore,

$$\kappa^{-1} = \frac{1}{B\sqrt{I}} = \frac{1}{B\sqrt{2c}} \tag{3.56}$$

and κ^{-1} will decrease by a factor of $1/\sqrt{2}$. Therefore, $\kappa^{-1} = \textbf{6.794x10}^{-9} \textbf{ m}$.

(c) Replacing 0.001 M NaCl by 0.001 M CdCl$_2$ will result in tripling the ionic strength ($I = 3c$) since CdCl$_2$ is a 2:1 electrolyte (cf. Exercise 3.11). Therefore, κ^{-1} will decrease by a factor of $1/\sqrt{3}$. Thus, $\kappa^{-1} = \textbf{5.547x10}^{-9} \textbf{ m}$

3.16 Estimate the dielectric constant of water in an aqueous solution of 5×10^{-4} M KCl at 25 ^{0}C using the Debye-Hückel limiting law. Comment on the result obtained. (Contractor)

Answer:

From the equation of activity coefficient in the *mksa* system (cf. Eq. 3.80 in the textbook), i.e.,

$$\ln f_{\pm} = -\frac{z_{+}z_{-}e_0^2 \kappa}{(4\pi\varepsilon_0)2\varepsilon kT} \tag{3.57}$$

the value of ε can be obtained if the values of $f_{\pm}$ and κ are known. The activity coefficient can be obtained from (cf. Eq. 3.90 in the textbook),

$$\log f_{\pm,KCl} = -A(z_+ z_-)\sqrt{I} \tag{3.58}$$

Since this is a 1:1 electrolyte, the ionic strength is given by $I = c$ (cf. Exercise 3.11), i.e., $I = c = 5 \times 10^{-4}$ M. Therefore,

$$\log f_{\pm,KCl} = -0.5115 \text{M}^{-1/2} \sqrt{5 \times 10^{-4} \text{ M}} = -0.0114$$
$$\Rightarrow f_{\pm,KCl} = 0.974 \tag{3.59}$$

The value of κ is obtained from the following equation expressed in the *mksa* system (cf. Eq. 3.19 in Exercise 3.6),

$$\kappa^{-1} = \left(\frac{\varepsilon \varepsilon_0 kT}{1000 e_0^2 N_A \sum_i c_i z_i^2} \right)^{1/2} = \left[\frac{\varepsilon \left(8.854 \times 10^{-12} \text{ C}^2 \text{J}^{-1} \text{m}^{-1} \right)}{\left(1000 \text{ dm}^3 \text{m}^{-3} \right) \left(1.602 \times 10^{-19} \text{ C} \right)^2} \right.$$

$$\times \left. \frac{\left(1.381 \times 10^{-23} \text{ JK}^{-1} \right) (298 \text{ K})}{\left(6.023 \times 10^{23} \text{ mol}^{-1} \right) \left(2 \times 5 \times 10^{-4} \text{ mol dm}^{-3} \right)} \right]^{1/2}$$

$$= \sqrt{\varepsilon} \left(1.535 \times 10^{-9} \text{ m} \right) \tag{3.60}$$

or

$$\kappa = \frac{\left(6.515 \times 10^8 \text{ m} \right)}{\sqrt{\varepsilon}} \tag{3.61}$$

Substituting Eqs. (3.59) and (3.61) into Eq. (3.57),

$$\ln 0.974 = -\frac{|1||-1| \left(1.602 \times 10^{-19} \text{ C} \right)^2 \left(6.515 \times 10^{-8} \text{ m} \right)}{\left(1.112 \times 10^{-10} \text{ C}^2 \text{J}^{-1} \text{m}^{-1} \right) (2) \left(1.381 \times 10^{-23} \text{ JK}^{-1} \right) (298) \varepsilon^{3/2}} \frac{1}{}$$

$$\tag{3.62}$$

Solving for ε gives,

$$\varepsilon = (702.9)^{2/3} = 79.05 \qquad (3.63)$$

Comment: The literature value for the dielectric constant of water is 78.3 at 298 K. The result shows the validity of the Debye-Hückel limiting law.

3.17 (a) Calculate the mean activity of a 0.0001 M KCl solution. (b) What is the activity change of KCl if 0.01 mol of ZnCl$_2$ is added to 1 liter of the above solution? (c) After the addition of the ZnCl$_2$ salt, will the Debye-Hückel reciprocal length increase or decrease? Explain. Consider $f_\pm = \gamma_{c\pm}$. (Cf. Exercise 3.30 in the textbook) (Xu)

Answer:

(a) The activity coefficient for a 1:1 electrolyte, f_{KCl}, is given by (cf. Eq. 3.91 in the textbook),

$$\log f_{\pm,KCl} = -A(z_+ z_-)\sqrt{I} = -A\sqrt{c}$$

$$= -0.5115 M^{-1/2} \sqrt{0.0001 M} = -5.12 \times 10^{-3} \qquad (3.64)$$

or

$$f_{\pm,KCl} = \gamma_{c\pm,KCl} = \mathbf{0.988} \qquad (3.65)$$

Therefore, the activity of KCl is,

$$a_{\pm,KCl} = c_{\pm,KCl} \gamma_{c\pm,KCl} = (0.0001)(0.988) = \mathbf{9.88 \times 10^{-5}} \qquad (3.66)$$

(b) After the addition of 1 mol ZnCl$_2$ to one liter of the above solution, the ionic strength is,

$$I = \frac{1}{2}\sum_i c_i z_i^2 = \frac{1}{2}\left[0.0001 M(1)^2 + 0.0001 M(1)^2 + 0.01(2)^2 + 0.02(1)^2\right]$$

$$= 0.0301 M \qquad (3.67)$$

Then,

$$\log f_{\pm,KCl} = -0.5115 M^{-1/2}(1)(1)\sqrt{0.0301 M} = -0.0887 \qquad (3.68)$$

or

$$f_{\pm,KCl} = \gamma_{c\pm,KCl} = 0.815 \qquad (3.69)$$

and the activity of KCl is,

$$a_{\pm,KCl} = c_{\pm,KCl}\gamma_{c\pm,KCl} = (0.0001)(0.815) = 8.15x10^{-5} \qquad (3.70)$$

(c) The Debye-Hückel reciprocal length, κ^{-1}, of KCl will decrease when $ZnCl_2$ is added to the solution. The increase of electrostatic force in the solution contracts the ionic atmosphere.

3.18 Calculate the Debye-Hückel reciprocal lengths for the following solutions at 298 K: 10^{-3} M NaCl, 0.1 M NaCl, 10^{-3} M $CaCl_2$, and 10^{-3} M $CaSO_4$. Use the constant values given in the textbook. (Cf. Exercise 3.23 in the textbook) (Kim)

Data:

$T = 298$ K $B = 0.3291x10^8 \, M^{-1/2} \, cm^{-1}$

Answer:

The Debye-Hückel reciprocal length, κ^{-1}, is given by (cf. Eq. 3.85 and Table 3.3 in the textbook),

$$\kappa^{-1} = \frac{1}{B\sqrt{I}} = \frac{1}{\left(0.3291x10^8 \, M^{-1/2} \, cm^{-1}\right)\sqrt{I}} \qquad (3.71)$$

NaCl is a 1:1 type electrolyte. Thus, for $c = 10^{-3}$ M, and considering the results from Exercise 3.11, $I = c = 10^{-3}$ M, and $\kappa^{-1} = 9.607x10^{-7}$ cm . In the same way for the other solutions,

	$z : z$	I	κ^{-1}
NaCl 10^{-3} M	1:1	$c = 10^{-3}$ M	$9.607x10^{-7}$ cm
NaCl 10^{-1} M	1:1	$c = 10^{-1}$ M	$9.607x10^{-8}$ cm
$CaCl_2$ 10^{-3} M	2:1	$3c = 3x10^{-3}$ M	$5.546x10^{-7}$ cm
$CaSO_4$ 10^{-3} M	2:2	$4c = 4x10^{-3}$ M	$4.803x10^{-7}$ cm

3.19 Calculate the highest concentration at which activity can be replaced by concentration in (a) NaCl, and (b) CaSO$_4$ solutions. Assume that a 10% error can be tolerated from the Debye-Hückel limiting law at 25 °C. Comment on these results. (Cf. Exercise 3.8 in the textbook) (Constantinescu)

Answer:

A permitted error of 10% in the activity means that the corresponding activity coefficient may be as low as 0.9. The Debye-Hückel limiting law establishes that (cf. Eq. 3.90 in textbook),

$$\log f_{\pm} = -A(z_+ z_-)I^{1/2} \tag{3.72}$$

The value of A at 25 °C is 0.5115 M$^{-1/2}$ and can be obtained from Eq. (3.89) or Table 3.4 in the textbook.

(a) For NaCl, a 1:1 valent aqueous electrolyte, $I = c$ (cf. Exercise 3.11). Thus, the highest allowed concentration is,

$$c = \left[-\frac{\log(0.9)}{0.5115\text{M}^{-1/2}\,|1||-1|} \right]^2 = 8.0\text{x}10^{-3}\ \text{M} \tag{3.73}$$

(b) For CaSO$_4$ that is a 2:2 valent aqueous electrolyte $I = 4c$ (cf. Exercise 3.11). Therefore, the highest allowed concentration is,

$$c = \frac{1}{4}\left[-\frac{\log(0.9)}{0.5115\text{M}^{-1/2}\,|2||-2|} \right]^2 = 1.3\text{x}10^{-4}\ \text{M} \tag{3.74}$$

Comment. The higher the valence of the electrolyte, the lower the limit of concentration at which activity can be replaced by the concentration.

3.20 Explain how the determination of the vapor pressure of water can give rise to the activity of the electrolyte dissolved therein. Start from the Gibbs-Duhem equation. (Cf. Exercise 3.17 in the textbook) (Bockris)

Answer:

The Gibbs-Duhem equation of thermodynamics establishes that,

$$\sum_i n_i d\mu_i = 0 \tag{3.75}$$

For a two components system, i.e., water and electrolyte,

$$n_w d\mu_w + n_e d\mu_e = 0 \tag{3.76}$$

The chemical potential of each component is given by,

$$\mu_i = \mu_i^0 + RT \ln a_i \qquad \text{or,} \qquad d\mu_i = RTd \ln a_i \tag{3.77}$$

where $d\mu_i$ is the chemical potential at standard conditions, and a_i is the activity of the component. Applying Eq. (3.74) to the electrolyte and water, and substituting into Eq. (3.76) gives,

$$n_w RTd \ln a_w + n_e RTd \ln a_e = 0 \tag{3.78}$$

or

$$d \ln a_e = -\frac{n_w}{n_e} d \ln a_w \tag{3.79}$$

Integrating the left-hand side from a low concentration of electrolyte, c_e, to a_e, and the right-hand side from ~ 1 to a_w,

$$\int_0^{a_e} d \ln a_e = - \int_1^{a_w} \frac{n_w}{n_e} d \ln a_w \tag{3.80}$$

Integrating the left-hand side of Eq. (3.80),

$$\ln \frac{a_e}{c_e} = - \int_1^{a_w} \frac{n_w}{n_e} d \ln a_w \tag{3.81}$$

The activity of the solvent can be written in terms of the vapor pressure of the pure solvent, $P_w{}^*$, and the vapor of the solvent when it is a component of a solution, P_w,

$$a_w = \frac{P_w}{P_w^*} \tag{3.82}$$

Therefore, writing Eq. (3.81) in terms of the corresponding vapor pressures of the solvent,

$$\ln \frac{a_e}{c_e} = - \int_1^{a_w} \frac{n_w}{n_e} d \ln \frac{P_w}{P_w^*} \tag{3.83}$$

The integral in Eq. (3.83) needs to be solved graphically. To do this, one plots n_w/n_e against $\ln P_w/P_w^*$, and the area will give the value of $\ln a_e/c_e$.

Comment: According to the limiting law, the plot of $f_{\pm}$ against $I^{1/2}$ should give a straight line of slope $-A\, z_-z_.$. Beside the valence of the ions constituting the particular electrolyte under consideration, the equation contains no reference to the specific properties of the salts that may be present in the solution.

Review of Section 3.5 of the Textbook.

Under what conditions does the Debye-Hückel limiting law apply? What are the inadequacies of this law? Explain how the theory was improved to allow for concentrated solutions of electrolyte. Following this theory, write equations for charge density, charge in the ionic cloud, potential at a distance r from the central ion taking into account the size of the ion, and the contribution from the ionic atmosphere to the potential. How are the individual and mean-ionic activities modified when the ion-size parameter, a, is considered? What range of values can a acquire? Can a be calculated through models? Discuss the success or failure of the introduction of the parameter a to the Debye-Hückel theory. Name the contribution of Milner and Gouy to the ionic solution theory.

3.21 Calculate the mean activity coefficients for 1:1, 1:2, and 2:2 valent electrolytes in water of ionic strengths 0.1, and 0.01 at 20 ^{0}C. The mean distance of closest approach of the ions is 3 Å. Use the extended Debye-Hückel limiting law and the corresponding constant values provided in the textbook. Comment on your results. (Cf. Exercise 3.15 in the textbook) (Constantinescu)

Data:

$T = 20\,^{0}C$ From Tables 3.3 and 3.4 in the textbook:
$a = 3\,Å$ $A = 0.5070\ \mathrm{M}^{-1/2}\ \ B = 0.3282 \times 10^{8}\ \mathrm{M}^{-1/2}\ \mathrm{cm}^{-1}$

Answer:

The extended Debye-Hückel limiting law establishes that (cf. Eq. 3.120 in the textbook):

$$\log f_{\pm} = -\frac{A\left(z_{+}z_{-}\right)I^{1/2}}{1 + BaI^{1/2}} \tag{3.84}$$

Therefore, for a 1:1 electrolyte,

$$\log f_{\pm} = -\frac{0.5070\,\mathrm{M}^{-1/2}\left|1\right|\left|-1\right|\left(0.1\mathrm{M}\right)^{1/2}}{1 + \left(0.3282 \times 10^{8}\ \mathrm{M}^{-1/2}\ \mathrm{cm}^{-1}\right)\left(3 \times 10^{-8}\ \mathrm{cm}\right)\left(0.1\mathrm{M}\right)^{1/2}} \tag{3.85}$$

$$= -0.1223$$

$$f_{\pm} = 0.7546 \tag{3.86}$$

Similarly,

Electrolyte	$I = 0.01\,\mathrm{M}$		$I = 0.1\,\mathrm{M}$	
	$\log f_{\pm}$	$f_{\pm}$	$\log f_{\pm}$	$f_{\pm}$
1:1	-0.0462	0.8992	-0.1223	0.7546
1:2	-0.0924	0.8083	-0.2446	0.5694
2:2	-0.1848	0.6534	-0.4892	0.3242

Comment: The mean-activity coefficient decreases with increasing ionic strength and increasing valence of the ions.

3.22 In Exercise 3.7, the Debye-Hückel reciprocal length, κ^{-1}, of 0.001 M NaCl solution was calculated as 9.607×10^{-9} m. Find the corresponding value of the size parameter, a, in the textbook. What conclusion related to the size of the ionic cloud could you draw from the comparison of these two values? (Kim)

Data:

$\kappa^{-1} = 9.607 \times 10^{-9}$ m $a = 0.40 \times 10^{-9}$ m (cf. Table 3.9 in the textbook)

Answer:

The difference between the equations of $f_\pm$ according to the point-charge model (cf. Eq. 3.90 in the textbook) and the finite-size model (cf. Eq. 3.121 in the textbook) is

$$\frac{1}{1 + a/\kappa^{-1}} \tag{3.87}$$

When this term tends to 1(i.e., when $\kappa^{-1} >> a$), then the equation for $\log f_\pm$ in the finite-ion-size model (cf. Eq. 3.121 in the textbook) approaches the corresponding equation for the point-charge model (cf. Eq. 3.90 in the textbook), i.e.,

$$\log f_\pm = -\frac{Az_+ z_- \sqrt{I}}{1 + a/\kappa^{-1}} \xrightarrow{\kappa^{-1} >> a} - Az_+ z_- \sqrt{I} \tag{3.88}$$

In this example, 10^{-3} M NaCl solution, the ratio

$$\frac{1}{1 + a/\kappa^{-1}} = \frac{1}{1 + \left(0.40 \times 10^{-9} \text{ m} / 9.607 \times 10^{-9} \text{ m}\right)} = 0.9897 \approx 1 \tag{3.89}$$

This means that, at this concentration, the ion atmosphere has such a large radius compared with that of the ion that one can ignore the finite size of the latter.

3.23 Calculate the product κa for the following solutions of CsCl: 10^{-4}, 5×10^{-4}, 10^{-3}, 5×10^{-3}, 10^{-2}, 5×10^{-2}, and 10^{-1} M at 25^0C. Find the maximum concentration at which $\kappa a < 0.1$, i.e., the concentration at which the limiting law is applicable. Assume that the distance of closest approach between two ions is equal to the sum of the ionic radii of the cation and the anion plus one diameter of water. (Cf. Exercise 3.28 in the textbook) (Bockris-GamboaAldeco)

Data:

$$a = r_{Cs^+} + r_{Cl^-} + 2r_w = 169 + 181 + 2(138) = 626 \text{ pm}$$
$$B = 0.3291 \text{ M}^{-1/2} \text{ cm}^{-1} \text{ (cf. Table 3.3 in the textbook)}$$

Answer:

The parameter κ is given by $\kappa = BI^{1/2}$ (cf. Eq. 3.85 in the textbook). Since CsCl is a 1:1 electrolyte, then, $I = c$ (cf. Exercise 3.11). Thus,

$$\kappa a = Ba\sqrt{c} = \left(0.3291 \text{M}^{-1/2} \text{cm}^{-1}\right)\left(6.26 \times 10^{-12} \text{cm}\right)\sqrt{c} \qquad (3.90)$$

For 10^{-4} M CsCl,

$$\kappa a = \left(0.3291 \times 10^8 \text{M}^{-1/2} \text{cm}^{-1}\right)\left(6.26 \times 10^{-8} \text{cm}\right)\sqrt{10^{-4}} \text{ M} = 0.0206 \ (3.91)$$

In the same way for the other concentrations,

c (M)	κa
1×10^{-4}	0.0206
5×10^{-4}	0.0461
1×10^{-3}	0.0652
5×10^{-3}	0.1457
1×10^{-2}	0.2060
5×10^{-2}	0.4607
1×10^{-1}	0.6215

The maximum concentration at which $\kappa a < 0.1$, and therefore the concentration at which the Debye-Hückel limiting law is applicable, is, then, about 1×10^{-3} M.

Review of Sections 3.6 to 3.8 of the Textbook.

Explain why the activity coefficient increases at high concentrations, i.e., at $c > \sim 1$N. Write an expression for the activity coefficient, $f_\pm$, to consider the solvent removal to the ions' aphere. Explain why the *unlinearized* Poisson-Boltzmann equation leads to logical inconsistency in the ion-solution theory. What is the difference between the Guntelberg charging process and the Debye charging process? What is an *ion-pair*? What is the physical meaning of the distance q according to the Bjerrum theory? Mention the main forces involved in the ion-pair formation as well as the main parameters that increase the occurrence of this phenomenon. How does ion-pair formation affect the Debye-Hückel theory?

3.24 Calculate the effect of water molecules on the activity coefficient in 1 M NaCl. The water activity in the solution is 0.96686, and the hydration number of the electrolyte is 3.5. The density of the solution is 1.02 g cm^{-3}. Consider the parameter $a = 4$ pm. (Cf. Exercise 3.11 in the textbook) (Kim)

Answer:

For a 1:1 electrolyte (cf. Eq. 3.130 in the textbook),

$$log f_{\pm} = -\frac{A\sqrt{c}}{1 + Ba\sqrt{c}} - 2.303\frac{n_h}{n}log\, a_w + 2.303 log\frac{n_w + n}{n_w + n - n_h} \qquad (3.92)$$

where n is the number of electrolyte moles in solution, n_w is the total number of water moles in solution, and n_s is the number of water moles in the solvation sheets of the ions. Thus, in 1 liter (or 1000 cm^3) of solution, $n = 1\,mol$, and

$$n_w = \frac{V_{soln}\,\rho_{soln} - n(MW)_{NaCl}}{(MW)_w}$$

$$= \frac{\left(1000\,cm^3\right)\left(1.02\,g\,cm^{-3}\right) - (1\,mol)\left(58.5g\,mol^{-1}\right)}{18.02g\,mol^{-1}} \qquad (3.93)$$

$$= 53.4\,mol$$

$$n_s = 3.5n = 3.5\,mol \qquad (3.94)$$

The activity coefficient *without* considering the effect of water, i.e., the first part of Eq. (3.92) is:

$$\log f_\pm = -\frac{A\sqrt{c}}{1 + Ba\sqrt{c}}$$

$$= -\frac{\left(0.5115 M^{-1/2}\right)\sqrt{1M}}{1 + \left(0.3291 \times 10^8 \, M^{-1/2} \, cm^{-1}\right)\left(0.4 \times 10^{-7} \, cm\right)\sqrt{1M}} \quad (3.95)$$

$$= -0.2208$$

or $\quad f_\pm = 0.6014$.

The terms related the hydration of the ions are, from Eq. (3.92):

$$-2.303 \frac{n_h}{n} \log a_w = -2.303 \frac{3.5 \, mol}{1 \, mol} \log 0.96686 = 0.1178 \quad (3.96)$$

$$+2.303 \log \frac{n_w + n}{n_w + n - n_h} = +2.303 \log \frac{53.4 \, mol + 1 \, mol}{53.4 \, mol + 1 \, mol - 3.5 \, mol} \quad (3.97)$$

$$= 0.0665$$

Adding these results to Eq. (3.95) gives,

$$\log f_\pm = -0.2208 + 0.1178 + 0.0665 = -0.0365 \quad (3.98)$$

or $f_\pm = 0.9194$. Therefore, the error when hydration is not considered is

$$\frac{0.9194 - 0.6014}{0.9194} \times 100 = 35\%.$$

3.25 According to the Bjerrum theory, at what concentration does KCl manifest 10% association in (a) ethanol, and (b) water? Use data obtained in tables in the textbook. (Cf. Exercise 3.12 in the textbook) (Bockris-GamboaAldeco)

Data:

$\theta = 10\%$ $\hspace{4cm}$ $|z_+| = |z_-| = 1$

From Tables 3.18 and 3.9 in the textbook:

$\varepsilon_{ethanol} = 24.30$ $\hspace{2.5cm}$ $\varepsilon_{water} = 78.3$ $\hspace{2.5cm}$ $a = 0.36 \, nm$

Answer:

(a) The concentration can be obtained from n_i^0 in Eq. (3.147) in the textbook written in the *mksa* system,

$$\theta = 4\pi n_i^0 \left[\frac{z_+ z_- e_0^2}{4\pi\varepsilon_0 \varepsilon kT} \right]^3 \int_2^b e^y y^{-4} \, dy \tag{3.99}$$

The parameter b in the above integral is obtained from Eq. (3.148) in the textbook. Thus, for the ethanol solution,

$$b = \frac{z_+ z_- e_0^2}{4\pi\varepsilon_0 a\varepsilon kT} = \frac{|1||-1|}{\left(1.112\times10^{-10}\ C^2 J^{-1} m^{-1}\right)\left(0.36\times10^{-9}\ m\right)(24.3)}$$

$$\times \frac{\left(1.602\times10^{-19}\ C\right)^2}{\left(1.381\times10^{-23}\ JK^{-1}\right)(298K)} = \left(\frac{2.307}{0.36\times10^{-9}\ m}\right) \tag{3.100}$$

$$= 6.41$$

The value of the integral in Eq. (3.99), can be found in Table 3.16 in the textbook. Thus, interpolating the value of $b = 6.41$ gives,

$$\int_2^b e^y y^{-4} \, dy = 1.409 \tag{3.101}$$

Therefore, n_i^0 from Eq. (3.99) is,

$$n_i^0 = \left[\frac{4\pi}{\theta}\left(\frac{z_+ z_- e_0^2}{4\pi\varepsilon_0 \varepsilon kT}\right)^3 \int_2^b e^y y^{-4} \, dy \right]^{-1} \tag{3.102}$$

$$= \left[\frac{4\pi}{0.1}\left(2.307\times10^{-9}\ m\right)^3 (1.409) \right]^{-1}$$

$$= 4.596\times10^{23}\ m^3$$

or,

$$\left(4.596 \times 10^{23} \text{ m}^{-3}\right) \times \frac{1 \text{m}^3}{1000 \text{dm}^3} \times \frac{1 \text{mol}}{6.022 \times 10^{23}} = 7.6 \times 10^{-4} \text{ M} \quad (3.103)$$

(b) In the same way, for water, with a dielectric constant of 78.3, $b = 1.989$, and Table 3.16 in the textbook indicates that $\int_2^b e^y y^{-4} dy = 0$. Therefore, **KCl does not reach 10% association in water** at any concentration.

Review of Sections 3.9 to 3.15 of the Textbook.

Mention the basis of the McMillan-Mayer theory for liquids. What type of catastrophe does one encounter when trying to calculate the energy of attraction between one central ion and ions surrounding it? What is the justification for adding the term $e^{-\kappa r}$ to the calculation of intermolecular energy by Mayer? Explain the Monte Carlo approach. Explain the difference between the Monte Carlo (MC) procedure and the Molecular Dynamic Simulation (MD). Mention some of the drawbacks of the computer-simulated procedures. Explain what a correlation function represents and mention some of its advantages. What is the Mean-Spherical approximation? Give some reasons for the discrepancy between theories of solutions and experimental data at high concentrations. Mention the contribution of Daves, Wertheim, and Haymet to the electrolyte theory. Under what conditions does Raman spectroscopy help detect the presence of ion pairs? Explain what is referred as "primitive" and "mound" model in the ionic solution theories. Describe different ways matter interacts with electromagnetic radiation. Write the Beer-Lambert law. Describe the Raman effect. What information about molecules can be obtained from Raman spectroscopy, infrared spectroscopy and nuclear magnetic resonance?

PROBLEMS

3.26 (a) When the Debye-Hückel model was developed, an important hypothesis was made for mathematical convenience, i.e., that $z_i e_0 \psi_r \ll kT$. Considering a 1:1 electrolyte solution at 25 °C as an example, reassess the validity of the hypothesis. (b) What is the physical nature of the above hypothesis? (Cf. Problem 3.2 in the textbook) (Xu)

Answer:

(a) Using the point charge version of Debye-Hückel model expressed in the *mksa* unit system (cf. Eq. 3.33 in the textbook),

$$\psi_r = \frac{z_i e_0}{4\pi\varepsilon_0\varepsilon} \frac{e^{-\kappa r}}{r} \tag{3.104}$$

Now, the approximation indicates that

$$z_i e_0 \psi_r \ll kT \tag{3.105}$$

Substituting the value of ψ_r and rearranging terms,

$$\frac{(z_i e_0)^2}{4\pi\varepsilon_0\varepsilon kT} \frac{e^{-\kappa r}}{r} \ll 1 \tag{3.106}$$

Consider now a 1:1 aqueous electrolyte at 25 °C, and a value of κ^{-1} in dilute solution of 10^{-8} m (cf. Table 3.2 in the textbook. It can be proved that κ^{-1} has a negligible effect on the results in diluted solutions.) The left-hand side (l.h.s.) of Eq. (3.106) as a function of the radius r becomes,

$$\frac{(z_i e_0)^2}{4\pi\varepsilon_0\varepsilon kT} \frac{e^{-\kappa r}}{r} = \frac{(1)^2 \left(1.602 \times 10^{-19}\ C \right)^2}{\left(1.112 \times 10^{-10}\ C^2 J^{-1} m^{-1} \right)(78.3)\left(1.381 \times 10^{-23}\ JK^{-1} \right)}$$

$$\times \frac{1}{(298\ K)} \frac{e^{-r/10^{-8}\ m}}{r} = \left(7.162 \times 10^{-10}\ m \right) \frac{e^{-r/10^{-8}\ m}}{r} \tag{3.107}$$

The hypothesis establishes according to Eq. (3.106) that,

$$\left(7.162 \times 10^{-10} \right) \frac{e^{-r\ 10^{-8}}}{r} \ll 1 \tag{3.108}$$

with r given in meters.

Evaluating the l.h.s. of Eq. (3.108) for different values of r, gives the results shown in the following table:

$r \times 10^8$ (m)	l.h.s. of Eq. (3.108)	$r \times 10^8$ (m)	l.h.s. of Eq. (3.108)
0.001	71.55	0.5	0.087
0.05	1.36	1	0.026
0.1	0.65	2	0.0049
0.2	0.29	3	0.0012
0.3	0.18	4	0.00033
0.4	0.12		

Plotting the l.h.s. of Eq. (3.104) against the distance from the center ion gives the graph in Fig. 3.2.

The validity of the hypothesis depends on the value of r. Close to the vicinity of the center of the ion, the hypothesis does not hold true, but as r increases, the function on the l.h.s. of Eq. (3.104) rapidly drops, and becomes much smaller than one. In Problem 3.28(a), it will be proved that the vast majority (more than 70%) of the ionic-cloud charge lies beyond a sphere of a radius κ^{-1}. In other words, more than 70% of the contribution to the ion atmosphere comes from a distance $r > \kappa^{-1}$. Under such conditions, the mathematical description employing the above hypothesis correctly reflects the average picture. However, as the solution becomes more concentrated, the value

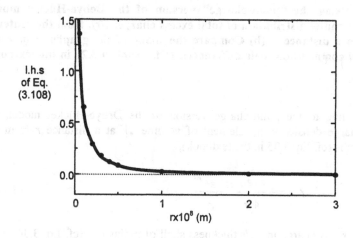

Figure 3.2. Plot of $\dfrac{(z_i e_0)^2}{4\pi\varepsilon_0 \varepsilon kT}\dfrac{e^{-\kappa r}}{r}$ in Eq. (3.107) vs. the distance from the central ion.

of κ^{-1} decreases (cf. Table 3.2 in the textbook), and the original hypothesis loses its validity. The breakpoint occurs at a concentration of about 10^{-3} M.

(b) To a first approximation in diluted solutions, when $r = \kappa^{-1}$, and $\kappa^{-1} = 10^{-8}$ m,

$$e^{-r/10^{-8}} = 0.37 \tag{3.109}$$

Taking $e^{-r/10^{-8}}$ as 0.37, Eq. (3.108) becomes,

$$\frac{2.64 \times 10^{-10} \text{ m}}{r} \ll 1 \tag{3.110}$$

or

$$r \gg 2.64 \times 10^{-10} \text{ m} \tag{3.111}$$

Therefore, the hypothesis adopted in the Debye-Hückel model is merely a dimensional demand: only the ionic species that are 2.64×10^{-10} m away from the center of the ion are considered in the model.

3.27 (a) Using the "point-charge" version of the Debye-Hückel model, derive the radial distribution of total excess charge, $q(r)$, from the center of the ion to a distance r. (b) Compare the shape of the graphs of $q(r)$ and $dq(r)$, and comment on their differences. (Cf. Problem 3.20 in the textbook) (Xu)

Answer:
(a) According to the point-charge version of the Debye-Hückel model, the excess charge density in an element of volume dV at a distance r from the central ion is (cf. Eq. 3.35 in the textbook),

$$\rho_r = -\frac{z_i e_0 \, \kappa^2 e^{-\kappa r}}{4\pi \quad r} \tag{3.112}$$

The excess charge in a dr-thickness shell of radius r is (cf. Eq. 3.36 in the textbook),

$$dq(r) = 4\pi r^2 \rho_r \, dr = -z_i e_0 \kappa^2 r e^{-\kappa r} \, dr \tag{3.113}$$

The total excess charge in the sphere of radius r is obtained by integrating Eq. (3.113) from $r = 0$ to $r = r$,

$$q(r) = \int_{r=0}^{r} dq(r) = -z_i e_0 \int_{r=0}^{r} (\kappa r) e^{-\kappa r} d(r\kappa) \qquad (3.114)$$

If $\kappa r = x$, then,

$$q(r) = -z_i e_0 \int_{r=0}^{r} xe^{-x} dx \qquad (3.115)$$

This integral can be solved by parts, i.e., $\int v du = vu - \int u dv$. Thus, considering $v = x$, $dv = dx$, $du = e^{-x} dx$, and $u = -e^{-x}$, then,

$$q(r) = -z_i e_0 \left(-xe^{-x} + \int_{r=0}^{r} e^{-x} dx \right) = -z_i e_0 \left(-xe^{-x} - e^{-x} \Big|_{r=0}^{r} \right)$$

$$= -z_i e_0 \left(-\kappa r e^{-\kappa r} - e^{-\kappa r} \Big|_{r=0}^{r} \right) \qquad (3.116)$$

or,

$$q(r) = -z_i e_0 \left(1 - \kappa r e^{-\kappa r} - e^{-\kappa r} \right) \qquad (3.117)$$

(b) Evaluating Eqs. (3.113) and (3.117), give the results shown in the next table:

κr	$-dq(r)/z_i e_0 \kappa dr$	$-q(r)/z_i e_0$
0	0	0
0.2	0.164	0.018
0.4	0.268	0.062
0.5	0.303	0.090
0.6	0.329	0.122
0.8	0.359	0.191
1	0.368	0.264
2	0.271	0.594
3	0.149	0.801
4	0.073	0.908
5	0.034	0.960

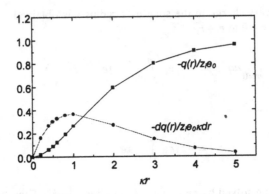

Figure 3.3. Comparison of $q(r)$ and $dq(r)$ functions.

Plots of $dq(r)$ and $q(r)$ functions against κr, give the graphs in Fig. 3.3. The amount of the total excess charge in the sphere of radius r around the ion, $q(r)$, increases monotonously with r. On the other hand, the amount of $dq(r)$ shows a maximum at $r = \kappa^{-1}$. When $r \to \infty$, $q(r)$ reaches its limiting value of $-z_i e_0$.

3.28 (a) Using the results from Problem 3.27, calculate the *total* excess charge within the sphere of the radius of the Debye-Hückel reciprocal length, κ^{-1}. How much of the overall excess charge has been accounted for within the sphere of radius κ^{-1}? **(b)** Plot $q(r)$ vs. r for an aqueous solution of 10^{-3} M 1:1 electrolyte at 25 °C. What distance from the ion encloses 95% of the excess charge? (Xu)

Answer:

(a) From Problem 3.27, Eq. (3.117), the excess charge from the center of the ion up to a radius r is

$$q(r) = -z_i e_0 \left(1 - \kappa r e^{-\kappa r} - e^{-\kappa r} \right) \tag{3.118}$$

Therefore, at $r = \kappa^{-1}$,

$$q(r) = -z_i e_0 \left(1 - e^{-1} - e^{-1} \right) = -0.264 z_i e_0 \tag{3.119}$$

A sphere of radius $r = \kappa^{-1}$ encloses only approximately a quarter of the total charge. Most of the excess charge is distributed out of the imaginary ionic cloud of radius κ^{-1}.

(b) From Table 3.2 in the textbook, κ^{-1} for an aqueous solution of 10^{-3} M 1:1 electrolyte at 25 °C is 9.6 nm. Thus, from Eq. (3.118),

$$q(r) = -z_i e_0 \left[1 - \left(1 + \frac{r}{9.6\,\text{nm}} \right) e^{-r/9.6\,\text{nm}} \right] \qquad (3.120)$$

Evaluating $q(r)/z_i e_0$ as a function of r,

r (nm)	$q(r)/z_i e_0$	r (nm)	$q(r)/z_i e_0$
1	0.005	40	0.936
5	0.096	45	0.948
9.6	0.264	50	0.966
20	0.616	60	0.986
30	0.819	70	0.994

Plotting $q(r)/z_i e_0$ as a function of r gives the graph in Fig. 3.4. The graph shows the radial distribution of total-excess charge, $q(r)/z_i e_0$, around the center ion. The limiting value of $q(r)$ at $r \to \infty$ is, without surprise, $-z_i e_0$. An excess charge of 95% is obtained at about 45 nm from the center of the ion.

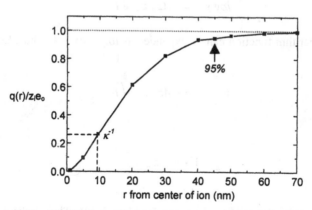

Figure 3.4. Total excess charge as a function of the distance from the central ion.

3.29 Find an equation relating the mean activity coefficient of thallous chloride with the molarity of this electrolyte in solutions containing KCl. The table below shows the solubility of TlCl in water and in the presence of various concentrations of potassium chloride solutions at 25 ^{0}C. Consider the activity of TlCl to be approximately constant in this range of concentration.

KCl concentration (mol kg^{-1})	0	0.025	0.050	0.10	0.20
TlCl solubility (mol kg^{-1})		1.607×10^{-2}	8.69×10^{-3}	5.90×10^{-3}	3.96×10^{-3}	2.68×10^{-3}

Consider that, for dilute solutions, $f_\pm \approx \gamma_\pm$ (cf. Problem 3.30). (Cf. Exercise 3.12 in the textbook) (Constantinescu)

Answer:

The mean activity coefficient is given by (cf. Eq. 3.55 in the textbook),

$$\gamma_\pm = \frac{a_\pm}{m_\pm} \qquad \text{or} \qquad \frac{1}{m_\pm} = \frac{\gamma_\pm}{a_\pm} \qquad (3.121)$$

and by (cf. Eq. 3.90 in the textbook),

$$log \gamma_\pm = -A z_+ z_- \sqrt{I} \qquad (3.122)$$

The logarithm function can be expanded as $log\, x \approx x - 1$. Thus, Eq. (3.122) becomes,

$$\gamma_\pm - 1 = -A z_+ z_- \sqrt{I} \qquad (3.123)$$

or

$$\gamma_\pm = 1 - A z_+ z_- \sqrt{I} \qquad (3.124)$$

For diluted solutions, $f_\pm \approx \gamma_\pm$ (see Problem 3.30). Thus, substituting Eq. (3.124) into Eq. (3.121),

$$\frac{1}{m_\pm} = \frac{1}{a_\pm} - \frac{Az_+z_-}{a_\pm}\sqrt{I} \tag{3.125}$$

This equation indicates that a plot of $1/m_\pm$ vs. $I^{1/2}$ should give a straight line with slope $Az_+z_-/a_\pm$ and intercept $1/a_\pm$, whenever $a_\pm$ is constant. From any of these parameters, it is possible to determine $a_\pm$. The different values of $m_\pm$ can be obtained from the definition of this variable, i.e.,

$$m_\pm = \left(m_+^{\nu_+} m_-^{\nu_-} \right)^{1/\nu} \tag{3.126}$$

For TlCl in pure water, $m_{Tl^+} = m_{Cl^-} = 1.607 \times 10^{-2}$ mol kg^{-1}, and $m_{\pm,TlCl}$ $= 1.607 \times 10^{-2}$ mol kg^{-1}. For the solution containing 0.025 mol kg^{-1} KCl, $m_{Cl^-} = 8.69 \times 10^{-3} + 0.025 = 33.7 \times 10^{-3}$ mol kg^{-1} and $m_{Tl^+} = 8.69 \times 10^{-3}$ mol kg^{-1}. Therefore,

$$m_{\pm,TlCl} = \sqrt{\left(8.69 \times 10^{-3}\right)\left(33.7 \times 10^{-3}\right)} = 17.1 \times 10^{-3} \text{ mol kg}^{-1} \tag{3.127}$$

The values of $m_\pm$ for the other solutions are given in the table below. The second parameter to determine is the ionic strength of the solutions. For the solution of TlCl in water, $I = c = 1.607 \times 10^{-2}$ mol kg^{-1}, (cf. Exercise 3.11). The ionic strength of the solution containing 0.025 mol kg^{-1} of KCl is,

$$I = \frac{1}{2}\left(m_{Tl^+} z_{Tl^+} + m_{Cl^-} z_{Cl^-} + m_{K^+} z_{K^+} \right)$$
$$= \frac{1}{2}(8.69 + 33.7 + 25) \times 10^{-3} = 33.7 \times 10^{-3} \text{ mol kg}^{-1} \tag{3.128}$$

In the same way for the other solutions,

m_{TlCl} (mol kg^{-1})	m_{KCl} (mol kg^{-1})	$m_\pm$ (mol kg^{-1})	I (mol kg^{-1})	$\gamma_\pm$
16.07x10^{-3}	0	16.07x10^{-3}	16.07x10^{-3}	0.8963
8.69x10^{-3}	0.025	17.1x10^{-3}	33.69x10^{-3}	0.8424
5.90x10^{-3}	0.050	18.2x10^{-3}	55.90x10^{-3}	0.7915
3.96x10^{-3}	0.100	20.2x10^{-3}	103.96x10^{-3}	0.7131
2.68x10^{-3}	0.200	23.3x10^{-3}	202.68x10^{-3}	0.6183

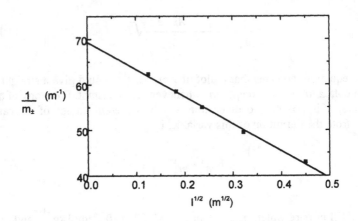

Figure 3.5. Mean activity coefficient of TlCl as a function of its concentration.

Plotting now $1/m_\pm$ vs. $\sqrt{I}$ gives the graph in Fig. 3.5. The fact that the graph gives a straight line validates Eq. (3.125). The equation corresponding to this graph is,

$$\frac{1}{m_\pm} = 69.41 - 59.88\sqrt{I} \tag{3.129}$$

Therefore, $a_\pm = 1/69.41 = 0.01441$, and from Eq. (3.121),

$$\gamma_{\pm, TlCl} = \frac{0.01441}{m_{\pm, TlCl}} \tag{3.130}$$

The last column in the above table gives the values of $\gamma_\pm$ at different concentrations of KCl.

3.30 The chemical potential of a given species in a real system, i.e., interacting particles, is given by:

$$\mu_i\left(real\right) = \mu_i^0 + RT\ln x_i f_i \tag{3.131}$$

where x_i is the concentration of the species i in mol-fraction units, μ_i^0 the chemical potential under standard state conditions and f_i is the activity coefficient when the concentration is expressed in mol fraction. However, the same equation can be written in terms of molality (m) or molarity (c) of species i as

$$\mu_i\left(real\right) = \mu_i^0\left(m\right) + RT\,ln\,m_i\gamma_{m,i} \tag{3.132}$$

$$\mu_i\left(real\right) = \mu_i^0\left(c\right) + RT\,ln\,c_i\gamma_{c,i} \tag{3.133}$$

Derive these two equations and find $\mu_i^0\left(m\right)$, $\mu_i^0\left(c\right)$, $\gamma_{m,i}$, and $\gamma_{c,i}$ in terms of μ_i^0 and f_i. (GamboaAldeco)

Answer:

(a) The molality is defined as the number of moles of solute (n) divided by the mass of the solvent, i.e., water, (n_w),

$$m = \frac{n}{m_w} = \frac{n}{n_w\,MW_w} \tag{3.134}$$

Dividing and multiplying by the total number of moles, i.e., $n_t = n + n_w$,

$$m = \frac{n}{n_w\,MW_w} \times \frac{n_t}{n_t} = \frac{x}{x_w\,MW_w} \tag{3.135}$$

where x and x_w are the mol fractions of the solute and water, respectively. Obtaining x from Eq. (3.135) and substituting it in the equation for $\mu_i(real)$, i.e., Eq. (3.131) gives,

$$\mu_i\left(real\right) = \mu_i^0 + RT\,ln\,mx_w\,MW_w\,f_i \tag{3.136}$$

The next step is to separate the argument of the logarithm and identify the different terms. However, to maintain the argument of the logarithms dimensionally correct, it is needed to multiply and divide by m^0, which is defined as $m^0 \equiv 1$ mol kg^{-1}. Thus,

$$\mu_i \left(real \right) = \mu_i^0 + RT \ln mx_w MW_w f_i \frac{m^0}{m^0}$$

$$= \mu_i^0 + RT \ln MW_w m^0 + RT \ln f_i x_w \frac{m}{m^0}$$

(3.137)

Defining

$$\mu_i^0 \left(m \right) \equiv \mu_i^0 + RT \ln \left(MW_w m^0 \right)$$

(3.138)

and

$$\gamma_{i,m} \equiv f_i x_w$$

(3.139)

gives,

$$\mu_i \left(real \right) = \mu_i^0 \left(m \right) + RT \ln \frac{m_i}{m^0} \gamma_{m,i}$$

(3.140)

or, when m_i is given in mol kg^{-1} (cf. Eq. 3.64 in the textbook),

$$\mu_i \left(real \right) = \mu_i^0 \left(m \right) + RT \ln m_i \gamma_{m,i}$$

(3.141)

Comment: For diluted solutions $x_w \rightarrow 1$. Then , from Eq. (3.139) $\gamma_{i,m} \rightarrow f_{i..}$

(b) The molarity is defined as the number of moles of solute, n, divided by the volume of the solution, V, i.e.,

$$c = \frac{n}{V}$$

(3.142)

Dividing and multiplying Eq. (3.142) by the total number of moles, n_t,

$$c = \frac{n}{V} \times \frac{n_t}{n_t} = \frac{x n_t}{V}$$

(3.143)

Obtaining x from Eq. (3.143) and substituting it in the equation for $\mu_i(real)$, i.e., Eq. (3.131), gives,

$$\mu_i\left(real\right)=\mu_i^0 +RT\ln c\overline{V}f_i \tag{3.144}$$

where $\overline{V}$ is the molar volume of the solution defined as $\overline{V} \equiv V/n_t$. Multiplying and dividing by c^0 defined as $c^0 \equiv 1\,mol\;dm^{-3}$,

$$\mu_i\left(real\right)=\mu_i^0 +RT\ln c\overline{V}f_i \,\frac{c^0}{c^0}=\mu_i^0 +RT\ln \overline{V}c^0 +RT\ln f_i \,\frac{c}{c^0} \tag{3.145}$$

Defining

$$\mu_i^0\left(c\right)\equiv \mu_i^0 +RT\ln \overline{V}c^0 \tag{3.146}$$

and

$$\gamma_{i,c} \equiv f_i \tag{3.147}$$

gives,

$$\mu_i\left(real\right)=\mu_i^0\left(c\right)+RT\ln c_i\gamma_{c,i} \tag{3.148}$$

when c_i is expressed in mol dm^{-3} (cf. Eq. 3.63 in the textbook).

3.31 Derive an expression for the solvent activity in diluted solutions. Hint: start with the Gibbs-Duhem equation considering the solution as a binary system. Then, apply the Debye-Hückel limiting law. (Cf. Exercise 2.21 in the textbook) (Xu)

Answer:

If the solution is considered a binary system, then system A represents the solvent (water), and system B the electrolyte. Thus, the Gibbs-Duhem equation reads (cf. Eq. 3.95 in the textbook):

$$n_A d\mu_A +n_B d\mu_B =0 \tag{3.149}$$

Now, the chemical potentials of the two systems are given by,

$$\mu_A =\mu_A^0 +RT\ln a_A \quad\text{and}\quad \mu_B =\mu_B^0 +RT\ln a_B \tag{3.150}$$

and their derivatives:

$$d\mu_A = RT \, d \ln a_A \quad \text{and} \quad d\mu_B = RT \, d \ln a_B \tag{3.151}$$

Therefore, Eq. (3.149) becomes,

$$n_A d \ln a_A = -n_B d \ln a_B \tag{3.152}$$

Since the solution is diluted, n_A can be regarded as a constant. Therefore, integrating Eq. (3.152),

$$\int_{a_A=1}^{a_A} d \ln a_A = - \int_{a_B=0}^{a_B} \frac{n_B}{n_A} d \ln a_B \tag{3.153}$$

or

$$\int_{a_A=1}^{a_A} d \ln a_A = - \int_{a_B=0}^{a_B} \frac{c_B}{c_A} d \ln a_B \tag{3.154}$$

The integral on the left side of Eq. (3.154) is simply

$$\int_{a_A=1}^{a_A} d \ln a_A = \ln a_A \tag{3.155}$$

The evaluation of the integral on the right-hand side of Eq. (3.154) is more complicated because a_B (or c_B) is a function of n_B. The activity of the electrolyte is defined as (cf. Eq. 3.55 in the textbook),

$$a_B = x_B f_B \tag{3.156}$$

where f_B the activity coefficient, and x_B is the concentration expressed in mole fraction, i.e.,

$$x_B = \frac{n_B}{n_B + n_A} = \frac{c_B}{c_B + c_A} \tag{3.157}$$

The next step is to find f_B, the activity coefficient of the electrolyte. This variable can be taken as the mean of the activity coefficients of the positive and negative ions constituting the electrolyte, i.e., $f_B = f_\pm$. Thus, considering the condition that the solution is very diluted and that the electrolyte is of the 1:1

type, the Debye-Hückel limiting law can be applied (cf. Eq. 3.90 in the textbook), i.e.,

$$\log f_B = \log f_{\pm} = -Az_+z_-\sqrt{I} = -A\sqrt{c_B} \tag{3.158}$$

where the condition $I = c_B$ for a 1:1 electrolyte was considered (cf. Exercise 3.11). Thus, taking logarithms on both sides of Eq. (3.156) and substituting the values of x_B and f_B, from Eqs. (3.157) and (3.158)

$$\ln a_B = \ln x_B + \ln f_B = \ln\frac{c_B}{c_B + c_A} - 2.303\,A\sqrt{c_B} \tag{3.159}$$

Since the solution is diluted, $c_B + c_A \approx c_A$, and

$$\ln a_B \approx \ln\frac{c_B}{c_A} - 2.303\,A\sqrt{c_B} \tag{3.160}$$

Differentiating Eq. (3.160),

$$d\ln a_B \approx \left(\frac{1}{c_B} - \frac{1.151\,A}{\sqrt{c_B}}\right)dc_B \tag{3.161}$$

Substituting Eqs. (3.155) and (3.161) into Eq. (3.154), and considering that $a_B = c_B$ gives

$$\ln a_A = -\int_0^{c_B}\frac{c_B}{c_A}\left(\frac{1}{c_B} - \frac{1.151\,A}{\sqrt{c_B}}\right)dc_B = -\frac{c_B}{c_A} + \frac{2}{3}\frac{1.151\,A\sqrt{(c_B)^3}}{c_A} \tag{3.162}$$

or

$$\ln a_A = \frac{0.767\,A\sqrt{(c_B)^3} - c_B}{c_A} \tag{3.163}$$

3.32 An electrochemical cell is represented by

$$Ag(s), AgBr(s) \mid KBr(aq) \parallel Cd(NO_3)_2\,(aq, 0.01\,M) \mid Cd(s) \tag{3.164}$$

The standard potentials on the hydrogen scale of the pairs Cd^{+2}/Cd and AgBr/Ag are –0.40 V and +0.07 V respectively. At 25 ^{0}C, the reading potential of the cell is – 0.62 V. Show that

$$\gamma_{\pm,KCl} = \frac{0.035}{c_{\pm,KCl}}$$
(3.165)

when the Debye-Hückel ion-cloud model holds well. (Contractor)

Answer:

The reactions on the right- and left-hand sides of the cell are:

r.h.s. $Cd^{+2}(aq) + 2e^- \rightarrow Cd(s)$ $E_1^0 = $ -0.40 V   (3.166)

l.h.s $2Ag(s) + 2Br^-(aq) \rightarrow 2AgBr(s) + 2e^-$ $E_1^0 = $ -0.07 V   (3.167)

$2Ag(s) + Cd^{+2}(aq) + 2Br^-(aq) \rightarrow 2AgBr(s) + Cd(s)$ $E_1^0 = $ -0.47 V   (3.168)

The measured potential is given by (cf. Eq. 3.103 in the textbook)

$$E = E^0 + \frac{RT}{nF} \ln\left(a_{Cd^{+2}} \, a_{Br^-}^2 \right)$$
(3.169)

where $E = $ -0.62 V, and $n = 2$. The value of $a_{Cd^{+2}}$ can be obtained from the activity coefficient equation, i.e. (cf. Eq. 3.90 in the textbook). Considering that $Cd(NO_3)_2$ is a 2:1 electrolyte, the ionic strength on the r.h.s of the cell is given by $I = 3c$ (cf. Exercise 3.11). Thus,

$$\log f_{\pm,Cd(NO_3)_2} = -A(z_+ z_-)\sqrt{I}$$
$$= -0.509\,M^{-1}\,|+2||-1|\sqrt{0.03\,M} = 0.18$$
(3.170)

or

$$f_{\pm,Cd(NO_3)_2} = 0.66$$
(3.171)

Since $f_\pm = \gamma_{c\pm}$ (cf. Problem 3.30),

$$a_{Cd^{+2}} = \gamma_{c\pm}c_{Cd^{+2}} = (0.66)(0.01) = 0.0067 \qquad (3.172)$$

Substituting the values of E, E^0, n, and $a_{Cd^{+2}}$ from Eq. (3.172) into Eq. (3.169),

$$-0.62\,V = -0.47\,V + \frac{\left(8.314\,J\,mol^{-1}K^{-1}\right)(298\,K)}{(2)\left(96500\,C\,mol^{-1}\right)} \times \frac{1\,CV}{1\,J}\ln\left(0.0067\,a_{Br^-}^2\right)$$

$$(3.173)$$

Solving for a_{Br^-},

$$a_{Br^-} = 0.035 \qquad (3.174)$$

Since $a_{Br^-} = \gamma_{c\pm}c_{KBr}$, therefore,

$$\gamma_{c\pm,KBr} = \frac{0.035}{c_{KBr}} \qquad (3.175)$$

3.33 The following electrochemical cell gives a potential of –0.218 V at 25 °C:

$$Ag \mid AgCl(s) \mid AgCl(aq.,sat.) \parallel AgCl(aq,sat.\,in\,0.05\,M\,CaCl_2) \mid AgCl(s) \mid Ag$$

$$(3.176)$$

The solubility product of AgCl is known to be 1.77×10^{-10} at room temperature. (a) Calculate the individual activity coefficient of Cl⁻ in the $CaCl_2$ solution. (b) Is there a way to obtain the individual activity coefficient for Ag^+ in the same solution? (Cf. Problem 3.6 in the textbook)(Xu)

Answer:

(a) The potential of the cell is related to the activities of the active species in both compartments of the cell by (cf. Section 3.4.8 in the textbook),

$$E_{cell} = E_{soln2} - E_{soln1} = \frac{RT}{F} \ln \frac{\left(a_{Ag^+}\right)_{right}}{\left(a_{Ag^+}\right)_{left}} \qquad (3.177)$$

The subscript *left* refers to the solution of AgCl in water (left side of the cell), and the subscript *right* to the solution of AgCl in 0.05 M CaCl$_2$ (right side of the cell).

On the left side of the cell, $AgCl(aq.,sat.)$ is in equilibrium with $AgCl(s)$. The solubility reaction of AgCl is:

$$Ag^+(aq.) + Cl^-(aq.) \rightarrow AgCl(s) \qquad (3.178)$$

and the corresponding solubility product (remember that the activity of pure substances is 1!):

$$K_s = \left(a_{Ag^+}\right)\left(a_{Cl^-}\right) = 1.77 \times 10^{-10} \qquad (3.179)$$

From Eq. (3.179),

$$\left(a_{Ag^+}\right)_{left} = \left(a_{Cl^-}\right)_{left} = \sqrt{1.77 \times 10^{-10}} = 1.33 \times 10^{-5} \text{ M} \qquad (3.180)$$

Substituting this value of $\left(a_{Ag^+}\right)_{left}$ into Eq. (3.177), and solving for $\left(a_{Ag^+}\right)_{right}$:

$$E_{cell} = \frac{\left(8.314 \text{ J mol}^{-1}\text{K}^{-1}\right)(298\text{K})}{96500 \text{ C mol}^{-1}} \ln \frac{\left(a_{Ag^+}\right)_{right}}{1.33 \times 10^{-5}} = -0.218 \text{ V} \quad (3.181)$$

$$\left(a_{Ag^+}\right)_{right} = 2.73 \times 10^{-9} \qquad (3.182)$$

Now, on the right side of the cell, $AgCl$ in 0.05 M $CaCl_2$ solution is in equilibrium with $AgCl(s)$. Neglecting the chloride from $AgCl$, the activity of chloride on the right side of the cell is,

$$K_s = \left(a_{Ag^+}\right)_{right}\left(a_{Cl^-}\right)_{right} = 2.73 \times 10^{-9} \left(a_{Cl^-}\right)_{right} = 1.77 \times 10^{-10} \quad (3.183)$$

$$\left(a_{Cl^-}\right)_{right} = \frac{1.77 \times 10^{-10}}{2.73 \times 10^{-9}} = 0.065 \quad (3.184)$$

The activity coefficient of chloride on the right side of the cell is given by

$$\left(\gamma_{Cl^-}\right)_{right} = \frac{\left(a_{Cl^-}\right)_{right}}{\left(c_{Cl^-}\right)_{right}} \quad (3.185)$$

With $\left(c_{Cl^-}\right)_{right} \approx 2 \times 0.05 = 0.10$,

$$\left(\gamma_{Cl^-}\right)_{right} = \frac{0.065}{0.10\,M} = 0.65 M^{-1} \quad (3.186)$$

(b) From the equation relating activity coefficient and ionic strength (cf. Eq. 3.90 in the textbook),

$$\log f_{\pm} = -A z_+ z_- I^{1/2} \quad (3.187)$$

The ionic strength in the right side of the cell neglecting the concentration of $AgCl$, is (cf. Eq. 3.83 in the textbook),

$$I = \frac{1}{2}\sum_i c_i z_i^2 = \frac{1}{2}\left(c_{Cl^-} + 4 c_{Ca^{+2}}\right)$$
$$= \frac{1}{2}[2(0.05) + 4(0.05)] = 0.15\,M \quad (3.188)$$

Considering that Eq. (3.187) holds for the right-hand side of the cell,

$$\left(\log f_{\pm}\right)_{AgCl,right} = -0.5115 M^{-1/2} (1)(1)\sqrt{0.15\,M} = -0.20 \quad (3.189)$$

or

$$\left(f_{\pm}\right)_{AgCl,right} = 0.63 \quad (3.190)$$

The individual activity coefficients of AgCl are related by,

$$f_{\pm AgCl} = \sqrt{f_{Ag^+} f_{Cl^-}}$$ (3.191)

Therefore,

$$0.63 = \sqrt{(f)_{Ag^+, right} (0.65)}$$ (3.192)

or

$$(f_\pm)_{Ag^+, right} = 0.61$$ (3.193)

3.34 (a) If the finite sized center ion has a size parameter a, what is the radial distribution of total excessive charge, $q(r)$ and $dq(r)$? **(b)** Use the *correspondance* principle to confirm the validity of this expression. **(c)** Does the Debye-Hückel reciprocal length, κ^{-1}, depend on a? Use the results from Problem 3.27. (Cf. Problem 3.10 in the textbook) (Xu)

Answer:

(a) The spatial distribution of the charge density is given by (cf. Eq. 3.34 in the textbook),

$$\rho_r = -\frac{\varepsilon}{4\pi}\kappa^2 \psi_r$$ (3.194)

If the ion is considered to have a finite size, a size parameter a can be assigned to it. Thus, the potential at a distance r from the finite-sized central ion is (cf. Eq. 3.113 in the textbook),

$$\psi_r = -\frac{z_i e_0}{\varepsilon} \frac{e^{\kappa a}}{1+\kappa a} \frac{e^{-\kappa r}}{r}$$ (3.195)

Substituting Eq. (3.195) into Eq. (3.194) gives,

$$\rho_r = -\frac{z_i e_0}{4\pi r} \frac{\kappa^2 e^{\kappa(a-r)}}{1+\kappa a}$$ (3.196)

The excess charge in a dr-thickness shell of radius r is (cf. Eq. 3.36 in the textbook),

$$dq(r) = 4\pi r^2 \rho_r\, dr \qquad (3.197)$$

Substituting ρ_r from Eq. (3.196) into Eq. (3.197) gives

$$dq(r) = -\frac{z_i e_0 \kappa^2}{1+\kappa a} re^{\kappa(a-r)}\, dr \qquad (3.198)$$

The total excess charge in the sphere of radius r is obtained by integrating Eq. (3.198) from a to a distance r. Therefore,

$$q(r) = \int_a^r dq(r) = -\frac{z_i e_0 e^{\kappa a}}{1+\kappa a} \int_a^r (\kappa r) e^{-\kappa r}\, d(r\kappa) \qquad (3.199)$$

Solving the integral by parts as in Problem 3.27,

$$q(r) = -\frac{z_i e_0 e^{\kappa a}}{1+\kappa a}\left(-\kappa r e^{-\kappa r} - e^{-\kappa r} \Big|_{r=a}^{r} \right) \qquad (3.200)$$

$$q(r) = -\frac{z_i e_0 e^{\kappa a}}{1+\kappa a}\left(-\kappa r e^{-\kappa r} - e^{-\kappa r} + \kappa a e^{-\kappa a} + e^{-\kappa a} \right) \qquad (3.201)$$

(b) To confirm the validity of this expression, we make use of the *correspondance principle*. The general version of a theory, in this case Eq. (3.201), must reduce to the approximate version under the conditions of applicability of the latter (cf. Section 3.5.3 in the textbook). In other words, does Eq. (3.201) reduces to Eq. (3.117) in Problem 3.27 under the appropriate conditions? Thus, when $a = 0$, Eq. (3.201) becomes,

$$q(r) = -z_i e_0 \left(-\kappa r e^{-\kappa r} - e^{-\kappa r} + 1 \right) \qquad (3.202)$$

This is the expression for the center ion as a point charge. This similitude validates the expression in Eq. (3.201).

(c) The parameter κ^{-1} represents the distance from the ion at which a spherical shell of infinitesimal thickness, dr, contains the maximum value of charge (cf. Sec. 3.3.8 in the textbook). Thus, κ^{-1} is obtained when the function dq has a maximum value, i.e., when $\dfrac{d}{dr}\dfrac{dq}{dr} = 0$. Thus, differentiating Eq. (3.198), and equating to zero,

$$\frac{d}{dr}\frac{dq(r)}{dr} = -\frac{z_i e_0 \kappa^2}{1+\kappa a}\frac{dq(r)}{dr} re^{\kappa(a-r)}$$

$$= -\frac{z_i e_0 \kappa^2}{1+\kappa a}\left[-r\kappa e^{\kappa(a-r)} + e^{\kappa(a-r)}\right] = 0 \tag{3.203}$$

Therefore,

$$r\kappa e^{\kappa(a-r)} = e^{\kappa(a-r)} \qquad \Rightarrow \qquad r = \kappa^{-1} \tag{3.204}$$

This indicates that, although the center ion is considered now to have a finite size, the distance at which $dq(r)$ has a maximum occurs still at κ^{-1}. In other words, the value of κ^{-1} **is independent of the ion size.**

3.35 Evaluate the Debye-Hückel constants A and B for ethyl alcohol. Use the calculated values to determine approximately the mean activity coefficients for 1:1, 1:2, and 2:2 valent electrolytes in this solvent at ionic strengths of 0.1 and 0.01 at 25 °C. The mean distance of closest approach of the ions, a, may be taken as 3 Å in each case. The dielectric constant of ethyl alcohol is 24.3. Comment on your results. (Cf. Problem 3.22 in the textbook) (Constantinescu)

Answer:

The two constants, A and B, are defined in the *mksa* unit system as (cf. Eqs. 3.86 and 3.89 in the textbook),

$$B = \left(\frac{2 N_A e_0^2}{\varepsilon \varepsilon_0 kT}\right)^{1/2} \qquad \text{and} \qquad A = \frac{1}{2.303}\frac{e_0^2}{4\pi\varepsilon_0 \varepsilon kT} B \tag{3.205}$$

Substituting the corresponding values for the parameter B,

$$B = \left[\frac{2\left(6.022 \times 10^{23}\ \mathrm{mol}^{-1} \right)\left(1.602 \times 10^{-19}\ \mathrm{C} \right)^2}{(24.3)\left(8.854 \times 10^{-12}\ \mathrm{C}^2 \mathrm{J}^{-1} \mathrm{m}^{-1} \right)\left(1.381 \times 10^{-23}\ \mathrm{JK}^{-1} \right)(298\,\mathrm{K})} \right]^{1/2}$$

$$= 1.869 \times 10^8\ \mathrm{mol}^{-1/2}\ \mathrm{m}^{1/2} \tag{3.206}$$

or (cf. Exercise 3.13),

$$B = \left(1.869 \times 10^8\ \mathrm{mol}^{-1/2}\ \mathrm{m}^{1/2} \right) \times \frac{\mathrm{dm}^{3/2}}{\mathrm{dm}^{3/2}} \left(\frac{1000\,\mathrm{dm}^3}{1\,\mathrm{m}^3} \right)^{1/2} \tag{3.207}$$

$$= 5.910 \times 10^9\ \mathrm{M}^{-1/2}\ \mathrm{m}^{-1}$$

The parameter A is given by,

$$A = \frac{\left(1.602 \times 10^{-19}\ \mathrm{C} \right)^2}{(2.303)(2)(24.3)\left(1.112 \times 10^{-10}\ \mathrm{C}^2 \mathrm{J}^{-1} \mathrm{m}^{-1} \right)\left(1.381 \times 10^{-23}\ \mathrm{JK}^{-1} \right)}$$

$$\times \frac{1.869 \times 10^8\ \mathrm{mol}^{-1/2}\ \mathrm{m}^{1/2}}{298\,\mathrm{K}} = 0.09365\,\mathrm{mol}^{-1/2}\ \mathrm{m}^{3/2} \tag{3.208}$$

or

$$A = \left(0.09365\,\mathrm{mol}^{-1/2}\ \mathrm{m}^{3/2} \right) \times \frac{\mathrm{dm}^{3/2}}{\mathrm{dm}^{3/2}} \left(\frac{1000\,\mathrm{dm}^3}{1\,\mathrm{m}^3} \right)^{1/2} = 2.961\,\mathrm{M}^{-1/2} \tag{3.209}$$

From the extended Debye-Hückel law (cf. Eq. 3.120 in the textbook), the equation for the mean activity coefficient reads,

$$\log f_{\pm} = -\frac{A z_+ z_- I^{1/2}}{1 + B a I^{1/2}} \tag{3.210}$$

For a 1:1 valent electrolyte in a solution with ionic strength of $I = 0.1$, Eq. (3.210) gives,

$$\log f_{\pm} = -\frac{\left(2.961\mathrm{M}^{-1/2}\right)(1)(1)(0.1\mathrm{M})^{1/2}}{1+\left(5.910\times10^{9}\ \mathrm{M}^{-1/2}\mathrm{m}^{-1}\right)\left(3\times10^{-10}\ \mathrm{m}\right)(0.1\mathrm{M})^{1/2}} \quad (3.211)$$

$$= -0.600$$

or

$$f_{\pm} = 0.251 \qquad\qquad (3.212)$$

In the same way for the other solutions,

Electrolyte	$I = 0.1$ M		$I = 0.01$ M	
	$\log f_{\pm}$	$f_{\pm}$	$\log f_{\pm}$	$f_{\pm}$
1:1	-0.600	0.251	-0.253	0.559
1:2	-1.200	0.0631	-0.506	0.312
2:2	-2.400	0.00398	-1.012	0.097

Comment. According to this theory, the mean activity coefficient decreases with increasing ionic strength of the solution. The decrement is greater as the valence of the ions increases. The unusual low values of mean activity coefficients obtained indicate the necessity to introduce other corrections that could reflect better the ion-ion interactions (cf. Chapter 3 in the textbook).

3.36 The radius of the ionic atmosphere is represented by κ^{-1}, **the Debye-Hückel length. When the solution is diluted, the ions are far apart, and the central ion sees a smoothed-out cloud of charge around itself. However, when the concentration increases, the distance,** ℓ**, between ions decreases, and one is confronting a situation in which the radius of the atmosphere is less than the average distance between ions. (a) Write an expression showing the average distance between ions,** ℓ**, as a function of the concentration,** c**. (b) Compare the distances** ℓ **and** κ^{-1} **for a low concentrated solution, e.g.,** 10^{-5} **M, and for a very concentrated solution, e.g.,** 10^{-2} **M, for a 1:1 electrolyte. (c) Find the concentrations for a 1:1 and a 2:2 electrolyte at which** $\ell > \kappa^{-1}$**, i.e., when the "disaster of coarse-**

grainedness" occurs. (d) Do you think an ionic-atmosphere model applies when $\ell > \kappa^{-1}$? (Cf. Problem 3.9 in the textbook) (Bockris-GamboaAldeco)

Answer:

(a) Consider the concentration c of a 1:1 electrolytic solution. In this solution, there are a total of $2c$ ions, or,

$$2c \, \text{mol dm}^{3} \times N_A \, \text{mol}^{-1} \times \frac{1000 \, \text{dm}^3}{1 \, \text{m}^3} = 2000 \, N_A c \text{ ions in 1 m}^3 \quad (3.213)$$

Therefore, the volume occupied by a single ion is, $1/2000N_A \, c \, \text{m}^3$. Now, to determine the distance between ions, ℓ, consider a model where the ions are in a cubic array, and each ion occupies a volume given by a cube of side ℓ, as shown in Fig. 3.6. In this array, the side of the cube is the same as the distance between the ions. Then, the volume occupied by one single ion is

$$V = \ell^3 = \frac{1}{2000 \, N_A c} \, \text{m}^3 \quad (3.214)$$

and the side of each cube in meters when the concentration is given in mol dm^{-3} is, then,

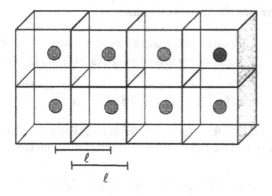

Figure 3.6. Cubic array of ions. The parameter ℓ represents the distance between ions.

$$\ell = \left(\frac{1}{2000 N_A c} m^3\right)^{1/3} = 9.398 \times 10^{-10} c^{-1/3} \qquad (3.215)$$

(b) The Debye-Hückel distance, κ^{-1} is given by the equation (cf. Eq. 3.85 in the textbook),

$$\kappa = B\sqrt{I} \qquad (3.216)$$

For a 1:1 electrolyte, $I = c$ (cf. Exercise 3.11). Therefore, κ^{-1} in meters is,

$$\kappa^{-1} = \left(0.3291 \times 10^8 \, M^{-1/2} cm^{-1} \sqrt{c} \times \frac{100 \, cm}{1 \, m}\right)^{-1} \qquad (3.217)$$

$$= 3.0386 \times 10^{-10} c^{-1/2}$$

Using Eq. (3.217), when $c = 10^{-5}$ M, then $\kappa^{-1} = 9.609 \times 10^{-8}$ m and $\ell = 4.3621 \times 10^{-8}$ m. Under this conditions, $\kappa^{-1} > \ell$. Thus, it is quite legitimate to argue that the central ion sees a *smoothed-out* cloud of charge around itself and one can use the Poisson equation with its implications of a continuous charge distribution. However, as the electrolyte concentration is increased, the situation ceases to be so satisfactory. For example, in a 10^{-2} M solution of the same electrolyte, $\kappa^{-1} = 3.0386 \times 10^{-9}$ m and $\ell = 4.3621 \times 10^{-9}$ m. At this concentration, $\kappa^{-1} < \ell$. The central ion experiences a *discrete* charge, not a smoothed-out cloud of charge. Thus, as the concentration increases, the ion cloud at a given distance from the ion, contains a number of ions which increases as the concentration increases. In other words, the cloud gets increasingly *coarse grained*. Smoothness decreases, discreteness increases.

(c) The "disaster of coarse grainedness" occurs when $\kappa^{-1} < \ell$. For a 1:1 electrolyte at 25 ^{0}C, ℓ and κ^{-1} are given by Eqs. (3.214) and (3.216). Thus,

$$3.0386 \times 10^{-10} c^{-1/2} < 9.398 \times 10^{-10} c^{-1/3} \qquad (3.218)$$

Solving Eq. (3.217) for c gives, $c > 0.0011$ M. For a 2:2 electrolyte $I = 2c$ (cf. Exercise 3.11), and the Debye-Hückel length given in meters is

$$\kappa = B\sqrt{I} = B\sqrt{2c} = 0.3291 \times 10^8 \sqrt{2}\sqrt{c} \qquad (3.219)$$

or

$$\kappa^{-1} = 2.148 x 10^{-10} c^{-1/2} \qquad (3.220)$$

From this equation and Eq. (3.216),

$$2.148 x 10^{-10} c^{-1/2} < 9.398 x 10^{-10} c^{-1/3} \qquad (3.221)$$

Simplifying terms shows that the disaster of coarse grainedness" for a 2:2 electrolyte occurs when $c > 0.00014$ M.

(d) No. The increase of grainedness leads to large fluctuations of the electrostatic potential, ψ, with time. Hence, the model of a continuous, smoothed-out charge density, i.e., the ionic cloud, must break down when this phenomenon occurs.

3.37 The experimental values of the mean activity coefficients for CaCl$_2$ at various concentrations at 25 ^{0}C are given in the Table below:

c (M)	$f_\pm$ (experimental)
0.0018	0.8588
0.0061	0.7745
0.0095	0.7361

Calculate the mean-effective-ion size, a. Calculate the mean-activity coefficient without considering and considering the ion size parameter, and comment on your results. (Contractor)

Answer:

For CaCl$_2$, $I = 3c$ (cf. Exercise 3.11). Thus, from the point-size model (cf. Eq. 3.91 in the textbook),

$$log f_\pm = - A z_+ z_- \sqrt{I} \qquad (3.222)$$

At 0.0018 M,

$$log f_\pm = -\left(0.5115 M^{-1/2}\right)|2||-1|\sqrt{3(0.0018 M)} = -0.07517 \qquad (3.223)$$

or

$$f_{\pm} = 0.8411 \qquad (3.224)$$

where $A = 0.5115$ M$^{-1/2}$ was taken from Table 3.4 in the textbook. At this concentration, the difference between $f_{\pm}$(experimental) and this calculated value is only 2%. This good agreement indicates that it is possible to determine the parameter a from this value of $f_{\pm}$(experimental) using the modified equation, and then recalculating the mean-activity-coefficient value for the other concentrations. Thus, from the finite-ion-size model (cf. Eq. 3.120 in the textbook),

$$\log f_{\pm} = -\frac{A z_+ z_- \sqrt{I}}{1 + Ba\sqrt{i}} \qquad (3.225)$$

Solving for a and substituting the corresponding values ($B = 0.3291 \times 10^8$ M$^{-1/2}$ cm^{-1} from Table 3.3 in the textbook),

$$a = -\frac{1}{B\sqrt{I}}\left(\frac{A z_+ z_- \sqrt{I}}{\log f_{\pm}} + 1\right) = -\frac{1}{0.3291 \times 10^8 \text{ M}^{-1/2}\text{cm}^{-1}\sqrt{3(0.0018\text{M})}}$$

$$\times \left[\frac{0.5115\text{M}^{-1/2}|2||-1|\sqrt{3(0.0018\text{M})}}{\log 0.8588} + 1\right]$$

$$= 5.671 \times 10^{-8} \text{ cm} \qquad (3.226)$$

With this value of a, $f_{\pm}$ is re-calculated at each concentration using the finite-ion-size model, i.e., Eq. (3.225),

$$\log f_{\pm} = -\frac{\left(0.5115 M^{-1/2}\right)|1||-1|\sqrt{3c}}{1 + \left(0.3291 \times 10^8 \text{ M}^{-1/2} \text{ cm}^{-1}\right)\left(5.671 \times 10^{-8} \text{ cm}\right)\sqrt{3c}} \qquad (3.227)$$

The next table shows $f_{\pm}$ for all the other concentrations using the point-size model as well as the finite-ion size model. The table includes the percentage difference between these values and the experimental ones, calculated as:

$$\%_{calc} = \frac{f_{\pm}(calc) - f_{\pm}(\exp)}{f_{\pm}(\exp)} \times 100 \qquad (3.228)$$

and

$$\%_{re-calc} \frac{f_\pm(re\text{-}calc) - f_\pm(\exp)}{f_\pm(\exp)} \times 100 \qquad (3.229)$$

c (M)	$f_\pm$ (exp)	$f_\pm$ point-size model	$\%_{calc}$	$f_\pm$ finite-ion size model	$\%_{re-calc}$
0.0018	0.8588	0.8411	2.1%	0.8588	-
0.0061	0.7745	0.7271	6.1%	0.7755	0.1%
0.0095	0.7361	0.6718	8.7%	0.7391	0.4%

The finite-ion-size model gives a better agreement with experimental values, compared to the point-ion model.

3.38. (a) Find the maximum potential at which the Poisson-Boltzmann equation can be linearized in a 1:1 electrolyte at 25 °C. (b) After linearization, the charge density in the atmosphere is proportional to the potential. Can you see any fundamental objection to having a ρ not proportional to ψ? (Cf. Problem 3.1 in the textbook) (Bockris-GamboaAldeco)

Answer:

(a) The Debye-Hückel law is valid at low concentrations, up to ~0.01 N for 1:1 electrolytes in aqueous solutions (cf. Section 3.5.1 in the textbook.) At this concentration, $\kappa^{-1} = 3.04$ nm (cf. Table 3.2 in the textbook.) Substituting this value in the corresponding equation for ψ_{cloud}, expressed in the *mksa* system (cf. Eq. 3.49 in the textbook),

$$\psi_{cloud} = -\frac{z_i e_0}{4\pi\varepsilon_0\varepsilon\kappa^{-1}} = -\frac{1\left(1.602\times10^{-19}\,C\right)}{\left(1.112\times10^{-10}\,C^2J^{-1}m^{-1}\right)(78.3)\left(3.04\times10^{-9}\,m\right)}$$

$$\times\frac{1CV}{1J} = -0.00605\,V \qquad (3.230)$$

(b) A charge density proportional to ψ is consistent with the law of superposition of potentials. However, a ρ not proportional to ψ would be the result of a non-linear relationship between ρ and ψ (cf. Section 3.7 in the textbook), which implies an invalidity in the law of superposition of potentials.

3.39 (a) As alternative to the Bjerrum's model of ion dissociation, use the results of the Debye-Hückel model to develop a qualitative but facile method to predict ion-pair formation. (Hint: compare the Debye-Hückel reciprocal length, κ^{-1}, with the effective distance for ion formation, q). (b) In lithium batteries, usually non-aqueous solutions constituted by lithium salts dissolved in organic solvents are used as electrolytes. The dielectric constants of such solvents range between 5 to 10. Using this information, estimate if ion paring occurs in these batteries, and compare with similar calculations in water. (Cf. Problem 3.13 in the textbook)(Xu)

Answer:

(a) The Debye-Hückel reciprocal length, κ^{-1}, defines a territory within which there is a considerable amount of counter ions (cf. Section 3.3.8 in the textbook). On the other hand, the effective distance for ion-pair formation defined by q in the Bjerrum's model encloses a territory within which two ions of opposite charge may associate into an ion-pair (cf. Eq. 3.8.3 in the textbook). Therefore, a fast method to predict ion-pair formation would be comparing the parameters q and κ^{-1}. Thus, if $q > \kappa^{-1}$, ion pair formation occurs in solution, but if $q < \kappa^{-1}$, ion pair does not form.

(b) The effective distance for ion-pair formation, q, is given by the Bjerrum equation in the *mksa* unit system as (cf. Eq. 3.144 in the textbook),

$$q = \frac{z_+ z_- e_0^2}{2(4\pi\varepsilon_0)\varepsilon kT} \tag{3.231}$$

Considering a 1:1 electrolyte at 25 ^{0}C,

$$q = \frac{\left(1.602 \times 10^{-19} \text{ C}\right)^2}{2\left(1.112 \times 10^{-10} \text{ C}^2 \text{J}^{-1} \text{m}^{-1}\right)\left(1.381 \times 10^{-23} \text{ JK}^{-1}\right)(298\,\text{K})\varepsilon} \cdot \frac{1}{} \tag{3.232}$$

$$= \frac{2.804 \times 10^{-8} \text{ m}}{\varepsilon}$$

If the electrolyte is *water*, $\varepsilon \approx 78$, and $q = 3.6 \times 10^{-10}$ m. Under these circumstances in diluted solutions $\kappa^{-1} \approx 10^{-8}$ m (cf. $c < 10^{-3}$ M in Table 3.2 in the textbook), and thus, $q \ll \kappa^{-1}$. This means that counter ions are beyond the distance of ion-pair formation, and thus this phenomenon is unlikely to occur. Now, if the electrolyte has a dielectric constant of, say, 10, as in poly(tetraglycol-dimethylether) in lithium batteries,

$$\kappa^{-1} = \left(\frac{\varepsilon \varepsilon_0 kT}{e_0^2 \sum_i n_i^0 z_i^2} \right)^{1/2} = \left\{ \frac{(10)\left(8.854 \times 10^{-12} \ C^2 J^{-1} m^{-1} \right)}{\left(1.602 \times 10^{-19} \ C \right)^2} \right.$$

$$\left. \times \frac{\left(1.381 \times 10^{-23} \ JK^{-1} \right)(298 \, K)}{\left[\left(2 \times 10^{-3} \ mol \, dm^{-3} \right)\left(6.022 \times 10^{23} \ mol^{-1} \right)\dfrac{1000 \, dm^3}{1 m^3} \right]} \right\}^{1/2} \quad (3.233)$$

$$= 3.43 \times 10^{-9} \ m$$

From Eq. (3.232), $q = 5.61 \times 10^{-9}$ m when $\varepsilon = 10$. Thus, $q > \kappa^{-1}$. This indicates that most of the counter ions are within an effective distance, and ion-pair formation is likely to occur.

3.40 (a) Describe the correlation function, as applied to ionic solutions and state two entirely different types of methods of calculating correlation functions. **(b)** Describe the "point" of knowing this quantity and comment on the meaning of the statement: "From a knowledge of the correlation function, it is possible to *calculate* solution properties, which may, in turn, be compared with the results of experiment. Hence, the calculation of the correlation function is the aim of all new theoretical work on solutions." (Bockris-GamboaAldeco)

Answer:

(a) The correlation function is a measure of the effect of species j in solution in increasing or decreasing the number of i ions in a given small volume of solution. Two methods to determine correlation functions are: Method 1 to determine correlation functions: By computer simulation MonteCarlo or

Molecular Dynamic methods. Method 2 to determine correlation functions: By experiment, using X-rays and neutron diffraction measurements.

(b) The main advantage is that there are no models assumed in the experimental determination of the correlation functions. In the case of computer simulated methods, correlation functions are obtained by assuming a certain pair-interaction law. In any case, once the correlation functions are obtained, different properties of the studied system can be determined without the experiment being troublesome, faster, and at lower costs. Thus, in a very general although complex way, the solution properties are found in terms of correlation functions. Since there are no model assumptions involved in the experimental determination of correlation functions, this approach is the aim of the determination of properties in solution.

MICRO-RESEARCH PROBLEMS

3.41 (a) One could use the Debye-Hückel ionic atmosphere model to study how ions of opposite charges attract each other. Based on this model, derive the radial distribution of cation concentration (n_+) and anion concentration (n_-) around a central positive ion in a dilute aqueous solution of a 2:2 electrolyte. (b) Plot these distributions, and compare this model with the Bjerrum's model of ion association. Comment on the applicability of this model to the study of ion-association behavior. (c) Using the data in Table 3.2 in the text, compute the cation and anion concentrations at the Debye-Hückel reciprocal lengths at various concentrations. Explain the applicability of the expressions derived. (Cf. Micro-research 3.1 in the textbook) (Xu)

Answer:

(a) The individual cation or anion concentration in a volume element dV around a given ion, is given by the Boltzmann distribution law, i.e. (cf. Eq. 3.7 in the textbook),

$$n_i = n_i^0 \, exp\left(-\frac{U}{kT} \right) \tag{3.234}$$

where n_i^0 is the bulk concentration or the calculated concentration throughout the whole solution, and U is the change in potential energy of the i particles when their concentration in the volume element dV is changed from the bulk value n_i^0 to n_i. If only the electrostatic contribution to the potential U is considered, then, (cf. Eq. 3.8 in the textbook),

$$n_i = n_i^0 \, exp\left(-\frac{z_i e_0 \psi_r}{kT} \right) \tag{3.235}$$

where ψ_r is the electrostatic potential. Using the approximation that $z_i e_0 \psi_r \ll kT$,

$$exp\left(\frac{z_i e_0 \psi_r}{kT} \right) \approx 1 - \frac{z_i e_0 \psi_r}{kT} \tag{3.236}$$

For a 2:2 electrolyte, $n^0 = n_+^0 = n_-^0$ and the expression for cation and anion concentrations in the volume element dV becomes,

$$n_+ = n^0 \left(1 - \frac{2 e_0 \psi_r}{kT} \right) \quad \text{for cations} \tag{3.237}$$

$$n_- = n^0 \left(1 + \frac{2 e_0 \psi_r}{kT} \right) \quad \text{for anions} \tag{3.238}$$

The next step is to find an appropriate expression for the electrostatic potential. Since the solution is diluted, the expression of ψ_r for the point-charge model can be applied. Hence, (cf. Eq. 3.33 in the textbook),

$$\psi_r = \frac{z_i e_0}{\varepsilon} \frac{e^{-\kappa r}}{r} \tag{3.239}$$

where z_i is the valence of the central ion, i.e., +2. Thus, inserting ψ_r from Eq. (2.239) into Eqs. (2.237) and (2.238), gives, for cations,

$$n_+ = n^0 \left(1 - \frac{4 e_0^2}{\varepsilon kT} \frac{e^{-\kappa r}}{r} \right) \tag{3.240}$$

or

$$\frac{n_+}{n^0} \cong 1 - \frac{4 e_0^2}{\varepsilon kT} \frac{e^{-\kappa r}}{r} \tag{3.241}$$

and for anions,

$$n_- = n^0 \left(1 + \frac{4 e_0^2}{\varepsilon kT} \frac{e^{-\kappa r}}{r} \right) \tag{3.242}$$

or

$$\frac{n_-}{n^0} = 1 + \frac{4 e_0^2}{\varepsilon kT} \frac{e^{-\kappa r}}{r} \tag{3.243}$$

(b) Figure 3.7 shows the profile of n_+ / n^0 and n_- / n^0 against the distance r from the center ion, where κ^{-1} was taken to be approximately 30 nm (cf. Eqs. 2.242 and 2.243). The figure shows that, the closer to the central ion, the larger the deviation from the bulk value, n^0, is. For the cations, the deviation means depletion, and for the anions, it means accumulation. This result is in agreement with the Debye-Hückel's picture of ionic cloud.

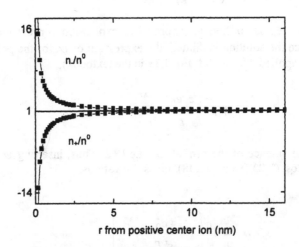

Figure 3.7. Radial distribution of cation and anion density around a positive center ion (+2). The deviation from the bulk value dies gradually with distance (hence the center ionic field).

In the Bjerrum's model of ion association, the same issue was handled, i.e., the distribution of anions and cations around a positive center ion [cf. Section 3.8 in the textbook]. While the initial reasoning is essentially the same, i.e., both theories consider an electrostatic distribution of ions, these two models are based on different approximations. The Bjerrum's model does not consider the ion-ion interaction when handling the electrostatic potential U responsible for the ion association, i.e., the contribution to U from other ions than the central one is ignored. This makes the Bjerrum model only good for diluted solutions.

On the other hand, the Debye-Hückel model considers inter-ionic effects, but to simplify its equations, it assumes that $z_i e_0 \psi_r \ll kT$, which holds only when r is large. In other words, the Debye-Hückel theory does not reflect a realistic view of ion distribution in the close neighborhood of the center ion. This disadvantage disqualifies this model for the study of the ion-association phenomenon, because the ion-pair formation only occurs in the close vicinity of the central ion. The advantage of this model is that it applies to larger concentrations than the Bjerrum's model.

(c) The cation and anion concentration at the Debye-Hückel reciprocal lengths for NaCl can be calculated from Eqs. (2.241) and (2.243). Thus, for 10^{-4} M, the cation concentration is,

$$
\frac{n_+}{n^0} = 1 - \left[\frac{4\left(1.602 \times 10^{-19}\ \mathrm{C}\right)^2}{(78.3)\left(1.381 \times 10^{-23}\ \mathrm{JK}^{-1}\right)(298\,\mathrm{K})\left(15.2 \times 10^{-9}\ \mathrm{m}\right)} \right.
$$

$$
\left. \times \frac{(0.3678)}{\left(1.112 \times 10^{-10}\ \mathrm{C}^2\,\mathrm{J}^{-1}\,\mathrm{m}^{-1}\right)} \right] = 0.9307
$$

(3.244)

which means that there is a small depletion of cations near the central ion. In the same way, the anion concentration is,

$$
\frac{n_-}{n^0} = 1 + \left[\frac{4\left(1.602 \times 10^{-19}\ \mathrm{C}\right)^2}{(78.3)\left(1.381 \times 10^{-23}\ \mathrm{JK}^{-1}\right)(298\,\mathrm{K})\left(15.2 \times 10^{-9}\ \mathrm{m}\right)} \right.
$$

(3.245)

$$\times \frac{(0.3678)}{\left(1.112 \times 10^{-10} \text{ C}^2 \text{J}^{-1} \text{m}^{-1}\right)} \Bigg] = 1.069$$

indicating that there is a small accumulation of anions at $r = \kappa^{-1}$. Similarly for higher concentrations,

Concentration (M)	κ^{-1} (nm)	n_+/n^0	n_-/n^0	$(z_i e_0)^2\, e^{-\kappa r} / \varepsilon r k T$
10^{-4}	15.2	0.9307	1.069	0.069
10^{-3}	4.81	0.7809	1.219	0.219
10^{-2}	1.52	0.3068	1.693	0.693
10^{-1}	0.48	(-1.195)	2.195	2.195

The result for n_+/n^0 when the concentration is 10^{-1} M makes no sense, because it gives a negative value and the concentration cannot be negative. To check the origin of this unreasonable result, one should look at the approximation used in the derivation of these equations. One may recall, that in expanding the exponent, $exp\,(-z_i e_0 \psi_i / kT)$, the hypothesis $z_i e_0 \psi_r \ll kT$ or $(z_i e_0)^2 e^{-\kappa r} / \varepsilon r k T \ll 1$, was made. The last column in the table shows that at 10^{-2} M the calculated value is already not much less than one, and at 10^{-1} M the value is larger than one. This indicates that at these high concentrations the hypothesis does not hold true, showing the reason for the appearance of negative concentrations. The theory is not valid any more.

CHAPTER 4

ION TRANSPORT IN SOLUTIONS

EXERCISES

Review of Sections 4.1 and 4.2 of the Textbook

Define *flux, diffusion, migration* and *hydrodynamic flow*. What forces are responsible for the movement of *mass* and *charge*, and the *diffusional flux* of species? Write equations defining these forces. What is *steady state*? Write *Fick's first law* of steady-state diffusion. In this equation, what does the parameter D represent? Is D a constant? Write expressions for the *average distance*, the *mean square distance*, and the *mean distance*. Write the *Einstein-Smoluchowski equation*, and explain its importance in diffusion. *Write Fick's second law* of time-dependent diffusion. How does this equation transform after applying Laplace transformation? Name three conditions applied to solve Fick's second law. Write the solution to the second Fick's law showing the space and time variation of concentration in response to a *constant-unit flux* that *extracts* ions from the system considered. How does this equation vary if now (a) the flux is *constant* but different from one, (b) the flux acts as a *source* of ions, (c) the flux varies as a *cosine* function, or (d) the flux varies as a *step* function? What is the *Einstein-Smoluchowski fraction*? What are the advantages and limitations of this equation? Write equations relating the *jump frequency* with the diffusion coefficient and with the activation-free energy. Briefly, describe two atomistic models of diffusion. Draw a diagram of the energy barrier for an ion jump. What is the *velocity-autocorrelation function*?

4.1 (a) Calculate the steady-state diffusional flux of Cl^- ions to the surface of an electrode if the concentration gradient is 0.5068 mol dm^{-3} cm^{-1}. (b) Calculate the magnitude of the diffusion current for the oxidation of chloride ions. Consider $D_{Cl^-} = 2.032 \times 10^{-5}$ cm^2 s^{-1}. (Contractor)

Answer:

(a) The steady-state diffusional flux is given by (cf. Eq. 4.16 in the textbook):

$$\overrightarrow{J_i} = -D_i \frac{dc_i}{dx} = -\left(2.032 \times 10^{-5} \text{ cm}^2 \text{s}^{-1} \right)\left(0.5068 \text{ mol dm}^{-3} \text{cm}^{-1} \right)$$
$$\times \frac{1 \text{ dm}}{10 \text{ cm}} \times \frac{100 \text{ dm}^2}{1 \text{ m}^2} = -1.030 \times 10^{-4} \text{ mol m}^{-2} \text{s}^{-1} \tag{4.1}$$

(b) The magnitude of the diffusion current is

$$i_D = zF \overrightarrow{J_i} = \left| -1 \right| \left(96500 \text{ C mol}^{-1} \right)\left(1.030 \times 10^{-4} \text{ mol m}^{-2} \text{s}^{-1} \right)$$
$$\times \frac{1 \text{ A}}{1 \text{ C s}^{-1}} = 9.940 \text{ A m}^{-2} \tag{4.2}$$

4.2 A certain radioisotope is leaking out from a defect in a storage tank. If the diffusion coefficient of the isotope is 3×10^{-5} cm^2 s^{-1}, calculate the mean distance traveled by it in 24 hours if transport occurs only by diffusion. (Contractor)

Answer:

From the Einstein-Smoluchowski equation (cf. Eq. 4.20 in the textbook),

$$\langle x \rangle = \sqrt{2Dt} = \left[2\left(3.0 \times 10^{-5} \text{ cm}^2 \text{ s}^{-1} \right)(24 \text{ hr}) \times \frac{3600 \text{ s}}{1 \text{ hr}} \right]^{1/2} = 2.2 \text{ cm} \tag{4.3}$$

4.3 In an instantaneous-pulse experiment, the electrode material is radioactive and hence, detectable by a Geiger counter. The pulse is produced with an electronic device generating a current of 10 A on a 0.1 cm^2 electrode during 0.1 second. A Geiger counter placed at 1 *cm* from the

electrode, registers the trace of the radioactive uni-valent ion at 450 seconds after the pulse. The diffusion coefficient of the released ions is of the order of 10^{-9} m^2 s^{-1}. (a) Draw two diagrams showing the experiment at $t = 0$ and at $t = 450$ s after the pulse. (b) Write an equation giving the quantity of ions produced during a single pulse, i.e., λ. (c) Determine λ for the conditions of the described experiment. (d) Calculate the limiting sensitivity of the instrument, that is, the minimum concentration of ions detected by the instrument. (Cf. Exercise 4.6 in the textbook) (Xu)

Data:

$I = 10$ A $d = 1$ cm $t_{pulse} = 0.1$ s
$Area = 0.1$ cm^2 $D = 10^{-9}$ m^2 s^{-1} $t = 450$ s

Answer:

(a) The diagrams showing the experiment at 0 s and 450 s are given in Fig. 4.1.

(b) The amount of the ion produced during a pulse is,

$$\lambda = \frac{I\,t_{pulse}}{FA} \tag{4.4}$$

where F is the Faraday constant and A the electrode area.

(c) Inserting the corresponding values in Eq. (4.4),

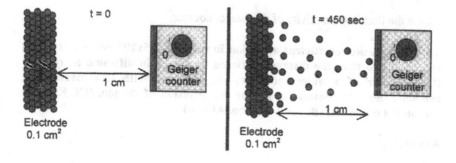

Figure 4.1. Schematics of the experiment described in Exercise 4.3 at $t = 0$ s and $t = 450$ s.

$$\lambda = \frac{(10\,A)(0.1s)}{\left(96500\,C\,mol^{-1}\right)\left(0.1cm^2\right)}\frac{1Cs^{-1}}{1A} = 1.036x10^{-4}\ mol\ cm^{-2}\quad (4.5)$$

(d) At the distance of the Geiger probe, i.e., $d = 1.0$ cm, the instrument detects radioactive material after 450 s. The concentration of the ion after the pulse at a distance x and after a time t is given by the equation (cf. Eq. 4.91 in the textbook),

$$c = \frac{\lambda}{\sqrt{\pi Dt}}\,exp\left(-\frac{x^2}{4Dt}\right)\quad (4.6)$$

Inserting the corresponding values in Eq. (4.6),

$$c = \frac{1.036\,mol\,m^{-2}}{\sqrt{\pi\left(10^{-9}\,m^2s^{-1}\right)(450s)}}\,exp\left(-\frac{(0.01m)^2}{4\left(10^{-9}\,m^2s^{-1}\right)(450s)}\right)\quad (4.7)$$

$$= 6.5x10^{-22}\ mol\ m^{-3}$$

or

$$c = 6.5x10^{-22}\ mol\ m^{-3}\times\frac{6.022\times10^{23}\ ions}{1\,mol}\times\frac{1m^3}{1000\,dm^3}\quad (4.8)$$

$$= 0.39\ ion\ dm^{-3}$$

This is the limiting sensitivity of the Geiger counter.

4.4 The diffusion coefficient of an ion in water is $1.5x10^{-5}$ cm^2 s^{-1}. It seems reasonable to take the distance between two steps in diffusion as roughly the diameter of a water molecule, i.e. 320 pm. With this assumption, calculate the rate constant in s^{-1}, for the diffusion of the ion. (Cf. Exercise 4.41 in the textbook) (Bockris-GamboaAldeco)

Answer:

The diffusion coefficient of the ion is related to the frequency of its jumps by (cf. Eq. 4.106 in the textbook),

$$D = \frac{1}{2} l^2 k \qquad (4.9)$$

and from this equation,

$$k = \frac{2D}{l^2} = \frac{1\left(1.5 \times 10^{-5} \text{ cm}^2 \text{ s}^{-1}\right)}{\left(320 \times 10^{-10} \text{ cm}\right)^2} = 1.5 \times 10^{10} \text{ s}^{-1} \qquad (4.10)$$

Review of Section 4.3 of the Textbook.

Define *potential, potential difference,* and *electric field.* What does the minus sign in the electric field equation indicate? Define *ionic conduction.* How does an electric field act on ions in solution? What force would oppose the applied electric field in an ionic solution? Explain how a steady flow of charge, i.e., current, is maintained in an electrolyte. Define *reduction* and *oxidation.* State *Kirckhoff's law* and *Faraday's law.* Write expressions for current density as a function of the flux of ions, $\vec{J}$, and the electric field, $\vec{X}$. Define *specific conductivity* and *resistance* of a solution. State *Ohm's law.* What are the conditions required so that an electrolytic conductor can be represented by Ohm's law? What is the relationship between *specific conductivity* and the *conductance* of an electrolytic solution? Define *molar conductivity* and *equivalent conductivity.* What is the definition and importance *of equivalent conductivity at infinite dilution,* and how is this quantity determined experimentally? State *Kohlrausch's law.* What is the convention adopted for the sign of the current flow? Do positive and negative ions in the same solution migrate independently?

4.5 A solution of 100 ml volume containing originally 10^{-2} M of Fe^{+3} ions is reduced at a constant current density of 100 mA cm^{-2} employing planar electrodes of 10 cm^2 area. Calculate the time after which the concentration of Fe^{+3} decreases by 10%. (Cf. Exercise 4.42 in the textbook) (Bock)

Data:

$V = 100 \text{ cm}^3$	$c^0_{Fe^{+3}} = 10^{-2} \text{ M}$	$\left(c_{Fe^{+3}}\right)_t = 0.9 \, c^0_{Fe^{+3}}$
$j = 100 \text{ mA cm}^{-2}$	$A_{electrode} = 10 \text{ cm}^2$	

Answer:

The current j passed through the circuit is related to the charge Q passed during the time t by

$$j = \frac{Q}{tA} \tag{4.11}$$

According to Faraday's law, the charge passed is also equivalent to

$$Q = nzF \tag{4.12}$$

where n is the number of Fe^{+3} ions converted into Fe^{+2}, i.e., $n = (c^0 - c)V = (1-0.9) c^0 V$, and z is the number of electrons involved in the reduction reaction of Fe^{+3}, i.e., $Fe^{+3} + 1e^- \rightarrow Fe^{+2}$, and thus, $z = 1$. Combining Eqs. (4.11) and (4.12) and substituting the corresponding values,

$$t = \frac{nzF}{jA} = \frac{(1-0.9)c^0 VzF}{jA} \tag{4.13}$$

$$= \frac{(0.1)\left(10^{-5} \, mol \, cm^{-3}\right)\left(100 \, cm^3\right)(1)}{\left(100 \times 10^{-3} \, A \, cm^{-2}\right)} \times \frac{6500 C \, eq^{-1}}{10 \, cm^2} \times \frac{1 A}{1 Cs^{-1}} = 9.6s$$

4.6 In the calibration of an amperometer, a coulometer of H_2 was connected in series in the circuit. After passing a current, 95.0 ml of H_2 were collected in one hour at 19 ^{0}C and 744 mm Hg. Calculate the current passed. (Tejada)

Data:

$V_{H2} = 95.0$ ml $T = 292$ K
$P_{H2} = 0.978$ atm $t = 3500$ s

Answer:

The number of moles of hydrogen collected are:

$$n = \frac{PV}{RT} = \frac{\left(0.978\,\text{atm}\right)\left(0.095\,\text{dm}^{-3}\right)}{\left(0.081\,\text{atm dm}^{-3}\text{mol}^{-1}\text{K}^{-1}\right)\left(292\,\text{K}\right)} = 3.92 \times 10^{-3}\ \text{mol} \quad (4.14)$$

From Eq. (4.13),

$$I = \frac{nzF}{t} = \frac{\left(3.92 \times 10^{-3}\ \text{mol}\right)\left(2\right)\left(96500\,\text{C mol}^{-1}\right)}{3600\,\text{s}} = 0.209\text{A} \quad (4.15)$$

4.7 A metallic wire transports a current of 1 A. How many electrons pass through one point of the wire per second? (Tejada)

Answer:

The current is given by the number of electrons per second that pass through the wire times the charge of the electrons, that is $I = N_e\, e_o$. Solving for N_e and substituting values, it is found that,

$$N_e = \frac{I}{e_o} = \frac{1\text{A}}{1.6022 \times 10^{-19}\ \text{C}} \times \frac{1\text{Cs}^{-1}}{1\text{A}} = 6.24 \times 10^{-18}\ \text{electrons s}^{-1} \quad (4.16)$$

4.8 Table 4.11 in the textbook lists data on equivalent conductivity measured at various concentrations. Calculate the conductance of these solutions if they were measured in a conductivity cell with 1 cm² electrodes separated 1 cm apart. Comment on the results obtained. (Contractor)

Answer:

The conductance is given by (cf. Eq. 4.134 in the textbook)

$$G = \sigma \frac{A}{l} \quad (4.17)$$

and the equivalent conductivity by (cf. Eq. 4.137 in the textbook)

$$\Lambda = \frac{\sigma}{cz} \qquad\qquad (4.18)$$

Combining Eqs. (4.17) and (4.18) to obtain the conductance as a function of the equivalent conductivity,

$$G = \Lambda cz\frac{A}{l} = \left(\Lambda\, S\, cm^2\, eq^{-1}\right)\left(cz\, eq\, dm^{-3}\right)(1cm)\times\frac{1\, dm^3}{1000\, cm^3} \qquad (4.19)$$

The obtained conductances are tabulated below:

Concentration (eq liter^{-1})	Λ (S cm^2 eq^{-1})	G (S)
0.001	146.9	1.469x10^{-4}
0.005	143.5	7.175x10^{-4}
0.01	141.2	14.12 x10^{-4}
0.02	138.2	27.64 x10^{-4}
0.05	133.3	66.65 x10^{-4}
0.1	128.9	128.9 x10^{-4}

Comment: The conductance of the solution increases as the concentration of the electrolyte increases. This is intuitively to be expected. However, the equivalent conductivity decreases with increasing concentration. Increasing ion-ion interaction with increasing ion population is responsible for this effect.

4.9 A conductance cell having a constant k (=l/A) of 2.485 cm^{-1} is filled with 0.01 N potassium chloride solution at 25 ^{0}C. The value of the equivalent conductivity for this solution is 141.2 S cm^2 eq^{-1}. If the specific conductivity, σ, of the water employed as solvent is 1.0x10^{-6} S cm^{-1}, what is the measured resistance of the cell containing the solution? (Cf. Exercise 4.18 in the textbook) (Constantinescu)

Data:

$k = 2.485$ cm^{-1} Λ_{soln} = 141.2 S cm^2 eq^{-1} $T = 25\ ^0$C

$c = 0.01$ N σ_{water} = 1.0x10^{-6} S cm^{-1}

Answer:

The specific conductivity is given by (cf. Eq. 4.134 in the textbook),

$$\sigma_{soln} = \frac{1}{R}\frac{l}{A} = \frac{k}{R} \qquad (4.20)$$

or,

$$R = \frac{k}{\sigma_{soln}} \qquad (4.21)$$

The value of σ_{soln} is also related to the equivalent conductivity by (cf. Eq. 4.137 in the textbook),

$$\sigma_{soln} = \Lambda_{soln}\, cz = \left(141.2\,\mathrm{S\,cm^2\,eq^{-1}}\right)\left(0.01\,\mathrm{eq\,dm^{-3}}\right)(1)\times\frac{1\,\mathrm{dm^3}}{1000\,\mathrm{cm^3}} \qquad (4.22)$$

$$= 1.412\times10^{-3}\,\mathrm{S\,cm^{-1}}$$

From Eq. (4.21),

$$R = \frac{2.485\,\mathrm{cm^{-1}}}{1.412\times10^{-3}\,\mathrm{S\,cm^{-1}}} = 1760\,\Omega \qquad (4.23)$$

Comment: Since $\sigma_{water} \ll \sigma_{soln}$, no correction is needed.

4.10 The specific conductivity of water at 298 K is $0.554\times10^{-7}\,\Omega^{-1}\,\mathrm{cm^{-1}}$. Calculate (a) the degree of dissociation and (b) the ionic product of water. Consider the molar conductivity of H^+ and OH^- to be $349.8\,\Omega^{-1}\,\mathrm{cm^2\,mol^{-1}}$ and $197.8\,\Omega^{-1}\,\mathrm{cm^2\,mol^{-1}}$ respectively. (Contractor).

Data:

$\sigma_w = 0.554\times10^{-7}\,\Omega^{-1}\,\mathrm{cm^{-1}}$ \qquad\qquad $\lambda_{m,H^+} = 349.8\,\Omega^{-1}\,\mathrm{cm^2\,mol^{-1}}$
$T = 298\,\mathrm{K}$ \qquad\qquad\qquad\qquad $\lambda_{m,OH^-} = 197.8\,\Omega^{-1}\,\mathrm{cm^2\,mol^{-1}}$

Answer:

(a) Let α be the degree of dissociation of the reaction

$$H^+ + OH^- \underset{\leftarrow}{\rightarrow} H_2O \qquad (4.24)$$

The concentrations of H^+ and OH^- in water are given by

$$c_{H^+} = c_{OH^-} = \alpha c_w \tag{4.25}$$

where c_w is the concentration of water, i.e., 55.56 M. Now, the specific conductivity of water is given by

$$\sigma_w = \sigma_{H^+} + \sigma_{OH^-} \tag{4.26}$$

and each individual-specific conductivity is related to the corresponding molar conductivity by (cf. Eq. 4.136 in the textbook) $\sigma_i = \lambda_{m,i} c_i$. Thus, Eq. (4.26) becomes,

$$\sigma_w = \lambda_{m,H^+} c_{H^+} + \lambda_{m,OH^-} c_{OH^-} \tag{4.27}$$

Substituting Eq. (4.25) into Eq. (4.27),

$$\sigma_w = \left(\lambda_{m,H^+} + \lambda_{m,OH^-}\right) \alpha c_w \tag{4.28}$$

The degree of dissociation is, then,

$$
\begin{aligned}
\alpha &= \frac{\sigma_w}{\lambda_{m,H^+} + \lambda_{m,OH^-}} \cdot \frac{1}{c_w} \\
&= \frac{0.554 \times 10^{-7}\,\mathrm{S\,cm^{-1}}}{(349.8 + 197.8)\mathrm{S\,cm^2\,mol^{-1}}} \cdot \frac{1}{55.56\,\mathrm{mol\,dm^{-3}}} \times \frac{1000\,\mathrm{cm}^3}{1\,\mathrm{dm}^3} \\
&= 1.822 \times 10^{-9}
\end{aligned}
\tag{4.29}
$$

(b) Taking into account Eq. (4.25), the dissociation constant of water is calculated as

$$
K_w = c_{H^+} c_{OH^-} = \alpha^2 c_w^2 = \left(1.822 \times 10^{-9}\right)^2 (55.56\,\mathrm{M})^2
$$
$$
= 1.023 \times 10^{-14}\,\mathrm{M}^2 \tag{4.30}
$$

4.11 The relation between resistance and specific conductivity, i.e., $R = l/\sigma A$ $= k/\sigma$ (cf. Eq. 4.134 in the textbook) gives a geometric definition of the cell constant k. However, different factors such as lack of ideal distribution of current lines and change of surface and distance parameter with constant use of the cell, might alter the value of the cell constant. Thus, the cell constant must be frequently re-determined by standardization with solutions of known specific conductivity. The most widely used solution is diluted KCl, for which internationally approved Reference Tables are available (ITS-90 Temperature Scale). With a given cell of plate surface 1.8 cm^2 and distance between plates of 0.8 cm, the following two conductance measurements at 298.15 K were performed: Sample 1: 0.01 M KCl, G = 2.845 mS [σ(ITS-90) = 1408.23 μS cm^{-1}]; Sample 2: 0.01 M K$_2$SO$_4$, G = 5.414 mS. (a) Calculate the specific conductivity and the equivalent conductivity of Sample 2. (b) Compare the ideal cell constant to the real one. (Trassati)

Data:

Sample 1: 0.01 M KCl, G = 2.845 mS [σ(ITS-90) = 1408.23 μS cm^{-1}]
Sample 2: 0.01 M K$_2$SO$_4$, G = 5.414 mS
A = 1.8 cm^2 l = 0.8 cm T = 298.15 K

Answer:

(a) From the standardization step given in Sample 1 the real cell constant can be calculated:

$$k = \frac{l}{A} = \frac{\sigma}{G} = \frac{1.40823\,\text{mS cm}^{-1}}{2.845\,\text{mS}} = 0.4950\,\text{cm}^{-1} \qquad (4.31)$$

The specific conductivity of Sample 2 is, then,

$$\sigma = G\frac{l}{A} = \left(5.414\,\text{mS}\right)\left(0.4950\,\text{cm}^{-1}\right) = 2.680\,\text{mS cm}^{-1} \qquad (4.32)$$

and the equivalent conductivity of Sample 2 is,

$$\Lambda = \frac{\sigma}{cz} = \frac{2.680\,\text{mS cm}^{-1}}{0.01\,\text{eq dm}^{-3}} \times \frac{1000\,\text{cm}^3}{1\,\text{dm}^3} = 268.0\ \text{S cm}^2\text{eq}^{-1} \qquad (4.33)$$

(b) The real cell constant as calculated in (a) is k_{real} = 0.4950 cm^{-1}, and the ideal cell constant is

$$k_{ideal} = \frac{l}{A} = \frac{0.80\,cm}{1.8\,cm^2} = 0.44cm^{-1} \qquad (4.34)$$

The percentage difference between k_{real} and k_{ideal} is, $[(0.4950 - 0.44)/0.44] \times 100 = 13\%$.

4.12 Use the data in Table 4.11 in the textbook to verify Kohlrausch's law. (Contractor)

Answer:

Kohlrausch's law indicates that there is a linear relationship between Λ and $c^{1/2}$ up to a concentration of ~0.01 N (cf. Section 4.3.9 in the textbook). The intercept on the y-axis of this curve gives Λ^0 (or Λ^∞), the equivalent conductivity at infinite dilution. The corresponding values of Λ at different concentrations are given in the next table. Figure 4.2 plots these results as well as the value of Λ^0.

c (eq liter^{-1})	$c^{1/2}$(eq liter^{-1})$^{1/2}$	Λ (S cm^2 eq^{-1})
0.001	0.03162	146.9
0.005	0.07071	143.5
0.01	0.1000	141.2
0.02	0.1414	138.2
0.05	0.2236	133.3
0.1	0.3162	128.9

The equation of the graph in Fig. 4.2 considering it as a straight line is $\Lambda =$ 149.5 – 0.79 $c^{1/2}$, with $\Lambda^\infty =$ 149.5 S cm^2 eq^{-1}. In this particular case, the law is valid for concentrations up to 0.02 eq dm^{-3}.

4.13 A conductance cell containing 0.01 M potassium chloride had a resistance of 2573 Ω at 25 °C. The same cell when filled with a solution of 0.2 N acetic acid had a resistance of 5085 Ω. Calculate (a) the cell constant, (k = l/A), (b) the resistivity of the potassium chloride and acetic acid solutions, and (c) the ratio of conductivities, $\alpha = \Lambda/\Lambda^0$, of the acetic acid

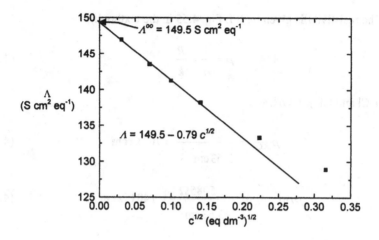

Figure 4.2. Verification of Kohlrausch's law

solution. At 25 °C, the equivalent conductivities at infinite dilution of the individual ions H^+ and CH_3COO^- are 349.82 and 40.9 Ω^{-1} cm^2 eq^{-1}, respectively. The specific conductivity of 0.01 N KCl solution is 0.0014114 Ω^{-1} cm^{-1}. (Cf. Exercise 4.16 in the textbook) (Constantinescu)

Data:

Solution 1: KCl	$c_1 = 0.01$ N	$R_1 = 2573\ \Omega$
Solution 2: CH_3COOH	$c_2 = 0.2$ N	$R_2 = 5085\ \Omega$
$\sigma(0.01$ N KCl$) = 0.0014114\ \Omega^{-1}\ cm^{-1}$		$T = 25\ °C$
$\lambda^0_{H^+} = 349.82\ \Omega^{-1}\ cm^2\ eq^{-1}$		$\lambda^0_{CH_3COO^-} = 40.9\ \Omega^{-1}\ cm^2\ eq^{-1}$

Answer:

(a) The specific conductivity is given by (cf. Eq. 4.134 in the textbook),

$$\sigma = \frac{l}{RA} = \frac{k}{R} \qquad (4.35)$$

where k is the cell constant. Thus,

$$k = R\sigma = (2573\Omega)\left(1.411 \times 10^{-3}\ \Omega^{-1}\ cm^{-1}\right) = 3.63\ cm^{-1} \qquad (4.36)$$

(b) The resistivity is given by (cf. Fig. 4.55 in the textbook),

$$\rho = \frac{1}{\sigma} = \frac{R}{k} \qquad (4.37)$$

For KCl and CH₃COOH,

$$\rho_{KCl} = \frac{2573\,\Omega}{3.63\,\mathrm{cm}^{-1}} = 709\,\Omega\,\mathrm{cm} \qquad (4.38)$$

$$\rho_{CH_3COOH} = \frac{5085\,\Omega}{3.63\,\mathrm{cm}^{-1}} = 1401\,\Omega\,\mathrm{cm} \qquad (4.39)$$

(c) The equivalent conductivity of the acetic acid solution is given by (Fig. 4.55 in the textbook),

$$\Lambda = \frac{\sigma}{cz} = \frac{1}{\rho cz} = \frac{1}{\left(1401\,\Omega\,\mathrm{cm}\right)\left(0.2\,\mathrm{eq}\,\mathrm{dm}^{-3}\right)(1)} \times \frac{10^3\,\mathrm{cm}^3}{1\,\mathrm{dm}^3} \qquad (4.40)$$

$$= 3.57\,\mathrm{S}\,\mathrm{cm}^2\,\mathrm{eq}^{-1}$$

The equivalent conductivity of the acetic acid solution at infinite dilution, Λ^0, is given by the law of independent migration of ions (cf. Eq. 4.144 in the textbook):

$$\Lambda^0_{CH_3COOH} = \lambda^0_{CH_3COO^-} + \lambda^0_{H^+}$$

$$= 349.82 + 40.9 = 390.7\,\mathrm{S}\,\mathrm{cm}^2\,\mathrm{eq}^{-1} \qquad (4.41)$$

Therefore, the ratio of conductivities for CH₃COOH is,

$$\frac{\Lambda}{\Lambda^0} = \frac{3.57\,\mathrm{S}\,\mathrm{cm}^2\,\mathrm{eq}^{-1}}{390.7\,\mathrm{S}\,\mathrm{cm}^2\,\mathrm{eq}^{-1}} = 0.00914 \qquad (4.42)$$

Review of Section 4.4 of the Textbook

What is the relationship between *drift velocity* and *absolute mobility*? Define *conventional mobility*. Write relationships between *conventional, specific, molar* and *equivalent* conductivity. What is the importance of these relationships? Write the *Einstein* equation. How is the diffusion coefficient related to the absolute mobility? Write *Stokes'* law. How is this law modified when instead of a spherical particle a cylindrical particle is considered? What is the main inconvenience of using Stokes' law in describing the movement of ions? Write the *Stokes-Einstein* and the *Nernst-Einstein* relations. Establish their importance in electrochemistry and the main factors that limit their validity. How does the diffusion coefficient vary with the concentration for a 1:1 electrolyte? What is a *phenomenological equation*? Write *Walden's rule*. When is the product $\Lambda \eta$ not a constant? Write equations for the *jump frequency* in terms of the *energy of activation*. What does the parameter β represent? Why is it said that the jumping frequency is *anisotropic* in the presence of a field? What is the relationship between the ionic current density and the electric field? How do *high-field* and *low-field Tafel* approximations apply to the current density equation? Write *Ohm's law*. Under what conditions does this equation apply? Define *electrochemical potential*. Write the *Nernst-Planck flux* equation and establish its importance in electrochemistry.

4.14 Calculate the absolute and the electrochemical (or conventional) mobilities of sodium ion when the drift velocity is 5.2×10^{-5} cm s^{-1} under an electrical field of 0.10 V cm^{-1}. (Cf. Exercise 4.29 in the textbook) (Kim)

Answer:

The absolute mobility is given by (cf. Eq. 4.149 in the textbook),

$$\bar{u}_{abs} = \frac{\vec{v}_d}{\vec{F}} = \frac{\vec{v}_d}{z_i e_0 \vec{X}} = \frac{5.2 \times 10^{-5} \text{ cm s}^{-1}}{|1| \left(1.602 \times 10^{-19} \text{ C}\right) \left(0.10 \text{ V cm}^{-1}\right)}$$

$$\times \frac{1 \text{ C V}}{1 \text{ J}} \times \frac{1 \text{ J}}{1 \text{ N m}} \times \frac{1 \text{ m}^2}{100 \text{ cm}^2} = 3.2 \times 10^{11} \text{ m N}^{-1} \text{s}^{-1} \qquad (4.43)$$

The electrochemical mobility is given by (cf. Eq. 4.152 in the textbook),

$$u_{conv} = \bar{u}_{abs} z_i e_0 = \frac{\vec{v}_d}{\vec{X}} = \frac{5.2 \times 10^{-5} \text{ cm s}^{-1}}{\left(0.10 \text{ V cm}^{-1}\right)} \times \frac{1 \text{ m}^2}{10^4 \text{ cm}^2} \qquad (4.44)$$

$$= 5.2 \times 10^{-10} \text{ m V}^{-1} \text{s}^{-1}$$

4.15 Calculate the conventional mobility of sodium ion in an aqueous solution knowing that the diffusion coefficient of the ion is 1.334×10^{-5} cm^2 s^{-1} at 25 ^{0}C. (Cf. Exercise 4.30 in the textbook) (Kim)

Answer:

From the Einstein equation, the absolute mobility is (cf. Eq. 4.172 in the textbook),

$$\bar{u}_{abs} = \frac{D}{kT} = \frac{1.334 \times 10^{-5} \text{ cm}^2 \text{ s}^{-1}}{\left(1.381 \times 10^{-23} \text{ J K}^{-1}\right)(298 \text{ K})} \times \frac{1 \text{ J}}{1 \text{ N m}} \times \frac{1 \text{ m}^2}{10^4 \text{ cm}^2} \qquad (4.45)$$

$$= 3.24 \times 10^{11} \text{ m N}^{-1} \text{s}^{-1}$$

and thus, the conventional mobility is (cf. Eq. 4.152 in the textbook),

$$u_{conv} = \bar{u}_{abs} z_i e_0 = \left(3.24 \times 10^{11} \text{ m s}^{-1} \text{N}^{-1}\right) |1| \left(1.602 \times 10^{-19} \text{ C}\right)$$

$$\times \frac{1 \text{ N m}}{1 \text{ J}} \times \frac{1 \text{ J}}{1 \text{ C V}} = 5.19 \times 10^{-8} \text{ m}^2 \text{V}^{-1} \text{s}^{-1} \qquad (4.46)$$

4.16 Calculate the radius of the solvated sodium ion in aqueous solution when the absolute mobility of the ion is 3.24×10^8 cm dyne^{-1} s^{-1}. The viscosity of the solution is 0.001 poise. (Cf. Exercise 4.31 in the textbook) (Kim)

Answer:

From Stoke's law (cf. Eq. 4.177 in the textbook),

$$r = \frac{1}{6 \pi \eta \bar{u}_{abs}} = \frac{1}{6 \pi (0.01 \text{ poise}) \left(3.24 \times 10^8 \text{ cm s}^{-1} \text{dyn}^{-1}\right)} \times \frac{1 \text{ poise}}{1 \text{ g cm}^{-1} \text{s}^{-1}} \qquad (4.47)$$

$$\times \frac{1 \text{g cm s}^{-2}}{1 \text{dyne}} \times \frac{1 \text{m}}{100 \text{cm}} = 1.64 \text{x} 10^{-10} \text{ m}$$

4.17 Calculate the equivalent conductivity of a 0.10 M NaCl solution. The diffusion coefficient of Na^+ is $1.334 \text{x} 10^{-5}$ cm^2 s^{-1} and that of the Cl^- is $2.032 \text{x} 10^{-5}$ cm^2 s^{-1}. (Cf. Exercise 4.20 in the textbook) (Kim)

Answer:

From the Nernst-Einstein equation (cf. Eq. 4.187 in the textbook),

$$\Lambda = \frac{zF^2}{RT}(D_+ + D_-)$$

$$= \frac{|1|\left(96500 \text{C mol}^{-1}\right)^2}{\left(8.314 \text{J K}^{-1} \text{mol}^{-1}\right)(298 \text{K})}\left(1.334 \text{x} 10^{-5} + 2.032 \text{x} 10^{-5}\right) \text{cm}^2 \text{s}^{-1} \quad (4.48)$$

$$\times \frac{1 \text{V}}{1 \text{A}} \times \frac{1 \text{A}}{1 \text{Cs}^{-1}} \times \frac{1 \text{J}}{1 \text{VC}} \times \frac{1 \text{m}^2}{10^4 \text{cm}^2} = 1.265 \text{x} 10^{-2} \text{ S m}^2 \text{eq}^{-1}$$

4.18 Calculate the conductivity of NaI in acetone. The viscosity of acetone is 0.00316 poise. The radii of the Na^+ and I^- ions are 260 and 300 pm, respectively. (Cf. Exercise 4.21 in the textbook)(Kim)

Answer:

The conductivity of the solution is given by the addition of the individual conductivities, i.e., (cf. Eq. 4.144 in the textbook),

$$\Lambda = \lambda_{Na^+} + \lambda_{I^-} \quad (4.49)$$

These individual conductivities can be calculated by Walden's rule (cf. Eq. 4.196 in the textbook), i.e.,

$$\lambda \eta = constant = \frac{ze_0 F}{6 \pi r} \quad (4.50)$$

Consider that,

$$\frac{1\,poise}{1\,g\,cm^{-1}s^{-1}} \times \frac{1\,V}{1\,A\Omega} \times \frac{1\,A}{1\,Cs^{-1}} \times \frac{1\,J}{1\,VC}$$

$$\times \frac{1\,kg\,m^2\,s^{-2}}{1\,J} \times \frac{1000\,g}{1\,kg} \times \frac{10^4\,cm^2}{1\,m^2} = \frac{1\,poise}{10^{-7}\,\Omega C^2\,cm^{-3}} \qquad (4.51)$$

Solving for λ_{Na^+} and substituting the corresponding values into Eq. (4.50),

$$\lambda_{Na^+} = \frac{|1|\left(1.602 \times 10^{-19}\,C\right)\left(96500\,C\,mol^{-1}\right)}{6\pi\left(260 \times 10^{-10}\,cm\right)(0.00316\,poise)} \times \frac{1\,poise}{10^{-7}\,\Omega C^2\,cm^{-3}} \qquad (4.52)$$

$$= 99.8\ S\,cm^2\,eq^{-1}$$

In the same way for iodide, $\lambda_{I^-} = 86.5\,S\,cm^2\,mol^{-1}$. The conductivity of NaI in acetone is, then, $\Lambda_{NaI} = 99.8 + 86.5 = 186.3\ S\,cm^2\,eq^{-1}$.

4.19 Walden's empirical rule states that the product of the equivalent conductivity and the viscosity of the solvent should be constant at a given temperature. (a) Explain the data in the table below obtained for NaI solutions in different solvents at 25 ^{0}C. (b) Calculate the radius of the moving entity in acetone applying Walden's rule. (Cf. Problem 4.10 in the textbook) (Bock)

Solvent	Ethanol	Acetone	Isobutanol
$\Lambda\eta\,(C^2cm^{-1}eq^{-1})$	5.247×10^{-8}	5.686×10^{-8}	4.883×10^{-8}
ε	24.3	20.70	15.80

Answer:

(a) Walden's rule is given by the equation (cf. Eq. 4.196 in the textbook),

$$\Lambda \eta = cons \tan t = \frac{ze_0 F}{6\pi r} \tag{4.53}$$

This equation shows that the product of the equivalent conductivity and the viscosity is a constant. However, this is subjected to the condition that the moving entity is the same independently of the solvent. Furthermore, the radius in the above equation enters through the Stokes-Einstein relation, which assumes that the moving entities are spherical ions (cf. Section 4.4.7 in the textbook). However, if there is a great difference between the dielectric constants of the solutions considered, then, several factors should be considered that may lead to deviations from the rule. On the one hand, different solvents produce different solvation of the ions, and thus, different radii of the moving entities. In addition, the degree of ion association increases with lowering the dielectric constant of the solvent, and thus, the concentration of free ions varies from solvent to solvent. In the three cases depicted in the table above, isobutanol has the lowest dielectric constant, and thus, it shows the largest deviation in the $\Lambda\eta$ product.

(b) From Walden's rule, i.e., Eq. (4.53),

$$r = \frac{ze_0 F}{6\pi\Lambda\eta} = \frac{|1|\left(1.6 \times 10^{-19}\,C\right)\left(96500\,C\,mol^{-1}\right)}{6\pi\left(5.686 \times 10^{-8}\,C^2\,cm^{-1}\,mol^{-1}\,eq^{-1}\right)} = 1.4 \times 10^{-8}\,cm \tag{4.54}$$

4.20 Calculate the concentration gradient of a univalent ion in 0.10 M solution at 25 °C when the electric field is 10^5 V cm^{-1}. (Cf. Exercise 4.2 in the textbook) (Kim)

Answer:

From the Nernst-Plank equation (cf. Eq. 4.226 in the textbook),

$$\vec{J} = \frac{DczF\,\vec{X}}{RT} - D\frac{dc}{dx} \tag{4.55}$$

When the flux is zero, $\vec{J} = 0$, and

$$\frac{dc}{dx} = \frac{czF\,\vec{X}}{RT} \tag{4.56}$$

Therefore, the concentration gradient is,

$$\frac{dc}{dx} = \frac{\left(0.10\,mol\,dm^{-3}\right)|1|\left(96500\,C\,mol^{-1}\right)\left(10^5\,V\,cm^{-1}\right)}{\left(8.314\,J\,mol^{-1}K^{-1}\right)\left(298\,K\right)} \times \frac{1J}{1CV} \quad (4.57)$$

$$= 3.9 \times 10^5 \, M\,cm^{-1}$$

Review of Section 4.5 of the Textbook

Define *transport number*. Mention the factors on which the conduction current of an ion depends. Write the modified *Nernst-Planck* equation for diffusion and diffusion flux. What is an *indifferent* or *supporting* electrolyte? What is the importance in electrochemical measurements of this type of electrolytes? What is *electroneutrality field*? What are *diffusion potential, concentration cell, liquid-junction potential* and *membrane potential*? How is the flux of one ionic species affected by the flux of the other species in solution? Describe three methods to measure transport numbers. What are the *Onsager-phenomenological equations*? What do L_{ij} and L_{ii} represent in these equations? Write an equation for diffusion potential as a function of transport numbers. Write the *Planck-Henderson* equation for diffusion. Is the transport number independent of concentration of the considered species?

4.21 A 0.2 N solution of sodium chloride was found to have a specific conductivity of 1.75×10^{-2} S cm^{-2} eq^{-1} at 18 ^{0}C, and a cation-transport number of 0.385. Calculate the equivalent conductivity of the sodium and chloride ions at this temperature. (Cf. Exercise 4.17 in the textbook) (Constantinescu)

Answer:

The transport numbers of the ions are related to the mobilities by (cf. Eq. 4.236 in the textbook),

$$t_i = \frac{u_{conv,i}}{\sum_i u_{conv,i}} \quad (4.58)$$

Since $u_{conv,i}$ is directly proportional to the equivalent conductivity (cf. Eq. 4.163 in the textbook) then,

$$t_i = \frac{\lambda_i}{\sum_i \lambda_i} = \frac{\lambda_i}{\Lambda} \qquad (4.59)$$

The equivalent conductivity of the solution is given by (cf. Eq. 4.163 in the textbook)

$$\Lambda = \frac{\sigma}{zc} = \frac{1.75 \times 10^{-2} \text{ S cm}^{-1}}{0.2 \text{ eq dm}^{-3}} \times \frac{1000 \text{ cm}^3}{1 \text{ dm}^3} = 87.5 \text{ S cm}^2 \text{eq}^{-1} \qquad (4.60)$$

Therefore, the equivalent conductivities of the individual ions are given by,

$$\lambda_{Na^+} = t_{Na^+} \Lambda = (0.385)\left(87.5 \text{ S cm}^2 \text{eq}^{-1} \right) = 33.69 \text{ S cm}^2 \text{eq}^{-1} \quad (4.61)$$

The equivalent conductivity of the chloride ion can be calculated by $\lambda_{Cl^-} = \Lambda - \lambda_{Cl}$ or by:

$$\lambda_{Cl^-} = t_{Cl^-} \Lambda = (1 - 0.385)\left(87.5 \text{ S cm}^2 \text{eq}^{-1} \right) = 53.81 \text{ S cm}^2 \text{eq}^{-1} \quad (4.62)$$

4.22 Estimate the concentration of indifferent electrolyte (e.g., K_2SO_4) which must be added to a 10^{-6} M HCl solution in order to study the diffusion of protons knowing that the cation mobilities at 25 °C for 0.1 M solutions are: $u_{conv,H^+} = 33.71 \times 10^{-4} \text{ cm}^2 \text{V}^{-1}\text{s}^{-1}$, **and** $u_{conv,K^+} = 6.84 \times 10^{-4} \text{ cm}^2 \text{V}^{-1}\text{s}^{-1}$. **(Cf. Exercise 4.11 in the textbook) (Bock)**

Answer:

There are four different species in solution, K^+, H^+, Cl^- and SO_4^{-2}. Their charges are $|z_{K+}| = |z_{H+}| = |z_{Cl-}| = 1$, and $|z_{SO4-2}| = 2$, and their concentrations, $c_{H+} = c_{Cl-} = 10^{-6}$ M and $c_{K+} = 2c_{SO4-2} = 2c_{K2SO4}$. The transport number of K^+ is, then, (cf. Eq. 4.234 in the textbook)

$$t_{K^+} = \frac{j_{K^+}}{j_{K^+} + j_{H^+} + j_{Cl^-} + j_{SO_4^{-2}}} = \frac{j_{K^+}}{\sum_i j_i}$$

$$= \frac{z_{K^+} F c_{K^+} u_{conv,K^+} \vec{X}}{\sum_i j_i}$$

(4.63)

Similarly,

$$t_{H^+} = \frac{j_{H^+}}{j_{K^+} + j_{H^+} + j_{Cl^-} + j_{SO_4^{-2}}} = \frac{j_{H^+}}{\sum_i j_i}$$

$$= \frac{z_{H^+} F c_{H^+} u_{conv,H^+} \vec{X}}{\sum_i j_i}$$

(4.64)

where (cf. Eq. 4.159 in the textbook) $j_i = z_i F c_i u_{conv,i} \vec{X}$. The role of the indifferent electrolyte is to carry the major part of the current. To assure that this is the case, the ratio of the transport number of K^+ to the transport number of H^+ should be large. Considering this number to be, say 200, then the ratio of t_{K^+}/t_{H^+} obtained from Eqs. (4.63) and (4.64), is,

$$\frac{t_{K^+}}{t_{H^+}} = \frac{z_{K^+} c_{K^+} u_{conv,K^+}}{z_{H^+} c_{H^+} u_{conv,H^+}} = \frac{2 c_{K_2 SO_4} \left(6.8 \times 10^{-4} \ cm^2 v^{-1} s^{-1}\right)}{10^{-6} \ M \left(33 \times 10^{-4} \ cm^2 v^{-1} s^{-1}\right)} = 200$$

(4.65)

Solving for $c_{K_2 SO_4}$,

$$c_{K_2 SO_4} = \frac{200 \left(10^{-6} \ M\right) \left(33 \times 10^{-4} \ cm^2 v^{-1} s^{-1}\right)}{2 \left(6.8 \times 10^{-4} \ cm^2 v^{-1} s^{-1}\right)} = 4.9 \times 10^{-4} \ M \quad (4.66)$$

4.23 A current of 5 mA flows through a 2 mm inner-diameter glass tube filled with 1 N $CuSO_4$ solution in the anode compartment and $Cu(CH_3SOO)_2$ solution in the cathode compartment. The interface created between the two solutions moves 6.05 mm towards the anode in 10 minutes. Calculate the transport number of the sulfate ion in this solution. (Cf. Exercise 4.27 in the textbook) (Herbert)

Data:

$I = 5$ mA	$x = 6.05$ mm	Anode: 1 N $CuSO_4$
$d = 2$ mm	$t = 10$ min	Cathode: $Cu(CH_3SOO)_2$

Answer:

This is the Lodge's experiment to determine transport numbers. The charge transported through the solution due to the migration of the sulfate ions equals the charge of such ions existing within the volume V limited by the initial and final position of the interface. Thus,

$$q_{SO_4^{-2}} = VFc \tag{4.67}$$

where $V = \pi r^2 x = \pi(0.1 \text{ cm})^2(0.605 \text{ cm}) = 0.0190 \text{ cm}^3$. The transport number is then, the ratio of $q_{SO_4^{-2}}$ to the total charge transported through the solution in the time given (cf. Eq. 4.262 in the textbook):

$$t_{SO_4^{-2}} = \frac{q_{SO_4^{-2}}}{It} = \frac{VFc}{It} = \frac{\left(0.0190 \times 10^{-3} \text{ dm}^3\right)\left(96500 C \text{ eq}^{-1}\right)}{(5 \text{mA})}$$

$$\times \frac{1 \text{eq dm}^{-3}}{10 \text{min}} \times \frac{1000 \text{mA}}{1 C s^{-1}} \times \frac{1 \text{min}}{60 s} = 0.611 \tag{4.68}$$

Comment: The mobile interface method illustrates an approach to determine the contribution of individual ionic species to conductivity.

4.24 Calculate the junction potentials for the following situations at 298 K: (a) 0.10 M HCl/0.01 M HCl, $t_+ = 0.83$, and (b) 0.10 M KCl / 0.01 M KCl, $t_+ = 0.49$. (Cf. Exercise 4.44 in the textbook) (Kim)

Answer:

The Planck-Henderson equation for liquid-junction potential considering a $z:z$ electrolyte is (cf. Eq.4.290) in the textbook),

$$-\Delta\psi = \frac{RT}{F}(t_+ - t_-)ln\left(\frac{c(l)}{c(0)}\right) \tag{4.69}$$

(a) For this solution, $t_+ = 0.83$, and thus, $t_- = 1 - 0.83 = 0.17$. Then,

$$-\Delta\psi = \frac{\left(8.314\,J\,mol^{-1}\,K^{-1}\right)(298\,K)}{96500\,C\,mol^{-1}}(0.83 - 0.17)ln\left(\frac{0.01}{0.10}\right) \times \frac{1C\,V}{1J} \tag{4.70}$$

$$= -0.039V$$

(b) For this solution, $t_+ = 0.49$, and thus, $t_- = 1 - 0.49 = 0.51$. Then,

$$-\Delta\psi = \frac{\left(8.314\,J\,mol^{-1}\,K^{-1}\right)(298\,K)}{96500\,C\,mol^{-1}}(0.49 - 0.51)ln\left(\frac{0.01}{0.10}\right) \times \frac{1C\,V}{1J} \tag{4.71}$$

$$= +0.0012V$$

Review of Section 4.6 of the Textbook

What is the *relaxation* approach? Why is it said that the ionic cloud is *egg* shaped? What are *relaxation field* and *electrophoretic force*? Draw a diagram of the forces acting on a moving ion. Write expressions for the *electrophoretic velocity, relaxation time, relaxation force, relaxation velocity,* and *drift velocity* on an ion. Write the *Debye-Hückel-Onsager* equation for a symmetric electrolyte. What is the importance of this equation? How would you improve this theory? What are the contributions of Justice, Fouss and Blum, and Lee and Wheaton to conductance theory?

4.25 Estimate the electrophoretic velocity of a sodium ion in 0.01 M NaCl solution under an electrical field of 0.1 V cm⁻¹. The viscosity of the solution is 0.00895 poise. (Cf. Exercise 4.3 in the textbook) (Kim)

Answer:

The electrophoretic velocity is given by (cf. Eq. 4.300 in the textbook),

$$v_E = \frac{ze_0}{6\pi\kappa^{-1}\eta} \vec{X} \tag{4.72}$$

The Debye-Hückel reciprocal length for a 1:1 electrolyte of concentration 0.01 M is (cf. Table 3.2) $\kappa^{-1} = 3.04$ mm. Substituting values into Eq. (4.72),

$$v_E = \frac{|1|\left(1.602 \times 10^{-19} \text{ C}\right)\left(0.1\text{V cm}^{-1}\right)}{6\pi\left(3.04 \times 10^{-7} \text{ cm}\right)(0.00895 \text{ poise})} \tag{4.73}$$

$$\times \frac{1 \text{poise}}{1 \text{g cm}^{-1} \text{s}^{-1}} \times \frac{1 \text{J}}{1 \text{VC}} \times \frac{10^7 \text{ g cm}^2 \text{s}^{-2}}{1 \text{J}} = 3.12 \times 10^{-6} \text{ cm s}^{-1}$$

4.26 Estimate the time for an ionic cloud to relax around a sodium ion in 0.1 M NaCl solution when the drift velocity is 5.2x10⁻⁵ cm s⁻¹ under an electrical field of 0.10 V cm⁻¹. (Cf. Exercise 4.39 in the textbook) (Kim)

Answer:

The relaxation time is obtained from (cf. Eq. 4.303 in the textbook):

$$\tau_R = \frac{\left(\kappa^{-1}\right)^2}{2\overline{u}_{abs} kT} \tag{4.74}$$

The absolute mobility is given by (cf. Eq. 4.149 in the textbook)

$$\overline{u}_{abs} = \frac{v_d}{\vec{F}} \tag{4.75}$$

The electrical force needed in Eq. (4.75) is obtained from

$$F = z_i e_0 \vec{X} = |1|\left(1.602 \times 10^{-19} \text{ C}\right)\left(0.10\text{V cm}^{-1}\right)$$

$$\times \frac{1 \text{J}}{1 \text{CV}} \times \frac{10^7 \text{ erg}}{1 \text{J}} \times \frac{1 \text{dyne cm}}{1 \text{erg}} = 1.6 \times 10^{-13} \text{ dynes} \tag{4.76}$$

Substituting Eq. (4.76) into Eq. (4.75),

$$\bar{u}_{abs} = \frac{5.2 \times 10^{-5} \text{ cm s}^{-1}}{1.6 \times 10^{-13} \text{ dynes}} = 3.25 \times 10^{8} \text{ cm s}^{-1} \text{ dynes}^{-1} \qquad (4.77)$$

The Debye-Hückel reciprocal length for a 1:1 electrolyte of this concentration is 0.96×10^{-9} m (cf. Table 3.2 in the textbook). Thus, the relaxation time becomes,

$$\tau_R = \frac{\left(0.96 \times 10^{-7} \text{ cm}\right)^2}{2\left(3.25 \times 10^{8} \text{ cm s}^{-1} \text{dyne}^{-1}\right)\left(1.381 \times 10^{-16} \text{ erg K}^{-1}\right)(298 \text{ K})} \qquad (4.78)$$

$$\times \frac{1 \text{ erg}}{1 \text{ dyne cm}} = 3.4 \times 10^{-10} \text{ s}$$

4.27 A rigorous treatment of the mobility of the ions considers three forces operating on the ion: (i) an electric force arising from the external field, (ii) a relaxation force from distortion of the cloud around a moving ion, and (iii) an electrophoretic force arising from the fact that the ion shares the electrophoretic motion of its ionic cloud. Considering these three forces, calculate the conventional mobility of a sodium ion in 0.01 M NaCl solution at 298 K. Take the viscosity of the solution as 0.00895 poise, the dielectric constant of the solution as 78.3, and the radius of the sodium ion as 260 pm. (Cf. Exercise 4.22 in the textbook) (Kim)

Answer:

The conventional mobility considering the three forces operating on the ion is given by (cf. Eq. 4.316 in the textbook),

$$u_{conv} = u^0 - \left(u_E + u_R\right) \qquad (4.79)$$

where u^0 is the result of the externally applied field only and excludes the influence of interactions between the ion and the ionic cloud. The terms u_E and u_R refer to the electrophoretic component of the mobility and to the relaxation field mobility, respectively. The term u^0 can be considered as the mobility at infinite dilution, given by the expression for the Stokes mobility, that is, (cf. Eq. 4.183 in the textbook)

$$u^0 = \frac{ze_0}{6\pi r \eta} = \frac{\left|1\right|\left(1.602 \times 10^{-19}\,C\right)}{6\pi\left(260 \times 10^{-12}\,m\right)(0.00895\text{poise})} \times \frac{1\,\text{poie}}{1\,g\,cm^{-1}s^{-1}} \times \frac{1\,J}{1\,C\,V}$$

$$\times \frac{1000g\,m^2\,s^{-1}}{1\,J} \times \frac{100cm}{1\,m} = 3.65 \times 10^{-4}\,cm^2\,V^{-1}s^{-1} \tag{4.80}$$

The electrophoretic component of the mobility is (cf. Eq. 4.316 in the textbook),

$$u_E = \frac{ze_0}{6\pi\eta\kappa^{-1}} \tag{4.81}$$

or

$$u_E = \frac{\left|1\right|\left(1.602 \times 10^{-19}\,C\right)}{6\pi(0.00895\text{poise})\left(3.04 \times 10^{-9}\,m\right)} \times \frac{1\,\text{poise}}{1\,g\,cm^{-1}s^{-1}}$$

$$\times \frac{1\,J}{1\,C\,V} \times \frac{1000g\,m^2\,s^{-1}}{1\,J} \times \frac{100cm}{1\,m} = 3.12 \times 10^{-5}\,cm^2\,V^{-1}s^{-1} \tag{4.81a}$$

where the Debye-Hückel reciprocal length for a 1:1 salt at a concentration 0.01 M is considered as 3.04×10^{-9} m (cf. Table 3.2 in the textbook). The relaxation component of the mobility given in the *mksa* system is (cf. Eq. 4.316 in the textbook),

$$u_R = \frac{1}{4\pi\varepsilon_0} \frac{u^0 e_0^2 w}{6\text{k}Tk^{-1}} = \frac{1}{1.112 \times 10^{-10}\,C^2\,J^{-1}m^{-1}}$$

$$\times \frac{\left(3.65 \times 10^{-4}\,cm^2\,V^{-1}s^{-1}\right)\left(1.602 \times 10^{-19}\,C\right)^2}{6(78.3)\left(1.38\text{lx}10^{-23}\,JK^{-1}\right)(298K)} \times \frac{0.5859}{3.04 \times 10^{-9}\,m} \tag{4.82}$$

$$= 7.165 \times 10^{-6}\,cm^2\,V^{-1}s^{-1}$$

$$u_R = \frac{1}{4\pi\varepsilon_0} \frac{u^0 e_0^2 w}{6 \&kT\kappa^{-1}} = \frac{1}{1.112\times10^{-10} \text{ C}^2\text{J}^{-1}\text{m}^{-1}}$$

$$\times \frac{\left(3.65\times10^{-4} \text{ cm}^2\text{V}^{-1}\text{s}^{-1}\right)\left(1.602\times10^{-19} \text{ C}\right)^2}{6(78.3)\left(1.381\times10^{-23} \text{ JK}^{-1}\right)(298\text{K})} \times \frac{0.5859}{3.04\times10^{-9} \text{ m}} \qquad (4.82)$$

$$= 7.165\times10^{-6} \text{ cm}^2\text{V}^{-1}\text{s}^{-1}$$

where $w = 0.5859$ for a 1:1 electrolyte is a correction factor introduced by Onsager (cf. Table 4.18 in the textbook). Then, substituting Eqs. (4.80)-(4.82) into Eq. (4.79) gives the conventional mobility,

$$u_{conv} = 3.65\times10^{-4} - \left(3.12\times10^{-5} + 7.165\times10^{-6}\right) \qquad (4.83)$$

$$= 3.26\times10^{-4} \text{ cm}^2\text{V}^{-1}\text{s}^{-1}$$

Review of Section 4.7 of the textbook. Define *relaxation time*. What is the physical meaning of this variable? Draw a schematic showing how to measure the electric conductance of ionic solutions by alternating currents of different frequencies. What is relaxation time of the asymmetry of the ionic cloud? What is the *Debye effect*? Describe the three parts in which the *dielectric constant* of water can be divided into according to *Kirkwood*. How does the dielectric constant vary with frequency? Explain. Describe the effects of ions on the relaxation times of the solvents. What is understood by *optical permitivity* and "*dielectrically saturated*"?

4.28 An investigator wants to study the Debye-effect of a diluted NaCl solution at room temperature, but he has no clue about what frequency range he should look at. Please help him. The diffusion coefficient of 0.001 M NaCl solution is 1.5×10^{-9} m^2 s^{-1}. (Cf. Exercise 4.43 in the textbook) (Xu)

Answer:

The Debye-Hückel reciprocal length for a 1:1 electrolyte of concentration 0.001 M at 298 K is, $\kappa^{-1} = 9.6\times10^{-9}$ m (cf. Table 3.2 in the textbook). According to the Einstein-Smoluchowski equation, the relaxation time of the ionic cloud is given by (cf. Eq. 4.27 in the textbook),

$$\tau = \frac{\left\langle x^2 \right\rangle}{2D} = \frac{\left(9.6 \times 10^{-9} \text{ m}\right)^2}{2\left(1.5 \times 10^{-9} \text{ m}^2 \text{s}^{-1}\right)} = 3.1 \times 10^{-8} \text{ s} \qquad (4.84)$$

Therefore, the corresponding frequency f at which the ion cloud relax is given by,

$$f = \frac{1}{2\pi\tau} = \frac{1}{2\left(3.1 \times 10^{-8} \text{ s}\right)} = 5.2 \times 10^{6} \text{ Hz} \qquad (4.85)$$

In order to observe the Debye effect, the applied frequency must be much higher than the above relaxation frequency, so that the relaxation of the ionic cloud cannot catch up with the applied field. Therefore, the investigator needs to look at the range above 10^7 Hz.

Review of Section 4.8 of the Textbook

Mention advantages of water as a solvent. Mention examples of industrial reactions where it is not possible to use water. Mention advantages and disadvantages of nonaqueous electrolytic solutions. Explain the parabolic-type curve observed in conductivity vs. concentration graphs for electrolytes in nonaqueous solvents. Explain the presence of temperature-dependent minima and maxima for salts in nonaqueous solvents. Mention relaxation processes related to changes of dielectric constant with frequency for nonaqueous solvents. What are the disadvantages of using Raman and NMR spectroscopies in the study of nonaqueous solvents? Mention advantages of liquid ammonia as solvent. Explain by the use of the *Debye-Hückel-Onsager* equation why the specific conductivities of nonaqueous electrolyte solutions are smaller than the specific conductivities of aqueous solutions at the same electrolyte concentrations. Explain via Bjerrum's theory the increase of ion association in nonaqueous solvents. What new dependence equivalent to Kohlrausch equation for aqueous electrolytes is applicable for nonaqueous electrolytes? What are *non-Coulombic* forces and what is their importance in *ion-pair* formation? What is a *triplet*? Under what conditions do these entities appear?

4.29 In acetonitrile, the equivalent conductivity of a very diluted solution of KI is 198.2 S cm^2 eq^{-1} at 25 ^{0}C. Calculate the equivalent conductivity of KI in a similar concentration range in acetophenone. The viscosity of

acetonitrile is 0.00345 poise, and that of acetophenone is 0.0028 poise. (Cf. Exercise 4.23 in the textbook) (Bockris-GamboaAldeco)

Data:

$\eta_{acetophenone} = 0.0028$ poise $\qquad \Lambda^0_{KI,acetonitrile} = 198.2 \ \text{S cm}^2 \text{eq}^{-1}$

$\eta_{acetonitrile} = 0.00345$ poise

Answer:

At infinite dilution (cf. Eq. 4.340 in the textbook),

$$u^0 \eta = cons\,tant \qquad (4.86)$$

The equivalent conductivity is related to the conventional mobility by (cf. Eq. 4.163 in the textbook) $\Lambda = Fu_{conv}$. Therefore, Eq. (4.86) becomes,

$$\Lambda^0 \eta = cons\,tant \qquad (4.87)$$

From the data of acetonitrile,

$$\Lambda^0_{acetonitrile} \eta_{acetonitrle} = \left(198.2 \text{S cm}^2 \text{eq}^{-1} \right) (0.00345 \text{poise})$$
$$= 0.684 \text{S poise cm}^2 \text{eq}^{-1} \qquad (4.88)$$

For acetophenone,

$$\Lambda^0_{acetophenone} = \frac{0.684 \text{S poise cm}^2 \text{eq}^{-1}}{0.0028 \text{poise}} = 244 \ \text{S cm}^2 \text{eq}^{-1} \qquad (4.89)$$

Comment: Walden's rule, i.e., Eq. (4.87), has to be carefully used. In the form presented in this problem, it is valid only under the assumption that the radii of the moving ions are the same in both solvents. This is the same as saying that the structures of their solvation sheets are similar in both solvents. A more generalized way for this rule is $u^0 r \eta$ = constant.

Review of Sections 4.9 to 4.11 of the Textbook

How does a polymer become a *conducting polymer*? What are the characteristics of conducting polymers? What is a *dopant*? Explain the conductivity in polymeric structures thorough *polaron* formation. What are *redox polymers, loaded ionomers,* and *electronically-conducting polymers*? Name applications of electronically conducting polymers in electrochemical science. Why is the proton considered a *different* sort of ion? What is meant by the conductance and mobility anomalies of the proton? Describe the *Grotthuss mechanism* for proton movement. Draw a *Morse* curve for the water-proton system. Write an equation for the frequency at which the proton crosses the potential barrier, i.e., $\vec{k}$. Is *Eyring's* theory on proton mobility in water successful in predicting the experimental values of mobility? What is *quantum-mechanical tunneling*? Is it reasonable to assume that proton can tunnel as the electron does? How did *Conway-Bockris-Linton* theory explain proton mobility? How does this theory explain the large mobility of protons in ice?

4.30 When it comes to practical applications, the actual conductance (the inverse of the resistance, R) instead of the specific conductivity is the important variable. This is the reason why polymer electrolytes have drawn so much attention as a potential player for alkali-metal batteries although their specific conductivities are usually low ($\sim 10^{-5}$ S cm^{-1}) compared with the non-aqueous electrolytes ($\sim 10^{-2}$ S cm^{-1}). Calculate the conductance of 1.0 M LiSO$_3$CF$_3$ in poly(ethylene oxide) and propylene carbonate, respectively. The former is fabricated into a film of thickness 10 μm, and the latter is soaked with porous separators of thickness 1 mm. (Cf. Exercise 4.15 in the textbook) (Xu)

Data:

$c_{LiSO3CF3} = 1.0$ M $\quad \sigma_{poly(ethylene\ oxide)} = 10^{-5}$ S cm^{-1} $\quad l_{poly(ethylene\ oxide)} = 10$ μm

$\sigma_{propylene\ carbonate} = 10^{-2}$ S cm^{-1} $\quad l_{propylene\ carbonate} = 1$ mm

Answer:

Considering an electrode of unit area, the conductance of the polymer electrolyte is [(cf. Eq. 4.134 in the textbook),

$$G_{poly\ (ethylene\ oxide)} = \sigma\frac{A}{l} = 10^{-5}\ S\,cm^{-1}\frac{1cm^2}{0.001cm} = 0.01\ S \qquad (4.90)$$

while the non-aqueous electrolyte has a conductance of

$$G_{propylene\ carbonate} = \sigma \frac{A}{l} = 10^{-2}\ S\,cm^{-1}\ \frac{1cm^2}{0.001cm} = 0.1\ S \qquad (4.91)$$

Comment: The good processability of polymer materials has partially made up for their inferior specific conductivity (1000 times smaller), reducing the gap in conductance to only 10 times. Other advantages of polymer electrolytes such as anti-leaking, safety, etc, make these materials serious candidates for industrial applications.

PROBLEMS

4.31 Does the valence of the ion affect the statement that the ionic-diffusion coefficient can be considered as a constant? To find out, take electrolytes of the $z{:}z$ type, for example 1:1 and 2:2, and compare the variation of their diffusion coefficients over the concentration range 0.01 - 0.1 mol dm^{-3}. (Cf. Problem 4.2 in the textbook) (Xu)

Answer:

The diffusion coefficient is given by (cf. Eq. 4.19 in the textbook):

$$D = BRT\left(1 + \frac{d\ln f_{\pm}}{d\ln c_i}\right) \qquad (4.92)$$

where the activity coefficient is given by the Debye-Hückel limiting law, i.e. (cf. Eq. 3.90 in the textbook),

$$\log f_{\pm} = -A z_+ z_- \sqrt{I} \qquad (4.93)$$

For $z{:}z$ type electrolytes, $|z_+| = |z_-| = z$ and $c_+ = c_- = c$. Thus, the ionic strength of this type of electrolytes is (cf. Eq. 3.83 in the textbook),

$$I = \frac{1}{2}\left(z_+^2 c_+ + z_-^2 c_-\right) = z^2 c \qquad (4.94)$$

Substituting Eq. (4.94) into Eq. (4.93),

$$\log f_{\pm} = -Az^3 \sqrt{c} \tag{4.95}$$

and Eq. (4.92) transforms into

$$D = BRT\left(1 - Az^3 \frac{d\sqrt{c}}{d\ln c}\right) \tag{4.96}$$

Combining the derivatives $d\sqrt{c}/dc = 1/2\sqrt{c}$, and $d\ln c/dc = 1/c$ gives, $d\sqrt{c}/d\ln c = c/2\sqrt{c} = \sqrt{c}/2$. Substituting this equation into Eq. (4.96),

$$D = BRT\left(1 - \frac{Az^3}{2}\sqrt{c}\right) \tag{4.97}$$

For an aqueous solution at 25 $^{\circ}$C, the parameter $A = 0.51$ M^{-1} (cf. Table 3.4 in the textbook). Thus, for a 1:1 electrolyte, the ratio of diffusion coefficients of the 0.01 M solution to the 0.1 M solution is,

$$\frac{D(0.1\text{M})}{D(0.01\text{M})} = \frac{1 - \left(0.255\text{M}^{-1}\right)(1)^3 \sqrt{0.1\text{M}}}{1 - \left(0.255\text{M}^{-1}\right)(1)^3 \sqrt{0.01\text{M}}} = 0.94 \tag{4.98}$$

and for a 2:2 electrolyte,

$$\frac{D(0.1\text{M})}{D(0.01\text{M})} = \frac{1 - \left(0.255\text{M}^{-1}\right)(2)^3 \sqrt{0.1\text{M}}}{1 - \left(0.255\text{M}^{-1}\right)(2)^3 \sqrt{0.01\text{M}}} = 0.45 \tag{4.99}$$

In the case of a 1:1 electrolyte there is only a ca. 5% change in its diffusion coefficient over the range of concentrations from 0.01 to 0.1 M. Hence, the diffusion coefficient of this type of electrolyte can be viewed as a constant. On the other hand, for a 2:2 electrolyte, there is a change of ca. 55%, and its diffusion coefficient cannot be considered constant in this range of concentrations.

4.30 (a) Derive and plot the relations for the variation of the ion concentration at the electrode surface ($x = 0$) under the conditions of (a) constant flux, and (b) instantaneous pulse. For the constant flux problem, consider fluxes of 1 and 10^{-1} mol cm^{-2} s^{-1}, with an initial concentration of 1 mol cm^{-3}, and a diffusion coefficient of 1.5×10^{-5} cm^{2} s^{-1}. For the instantaneously pulse problem, consider a total concentration of 10^{-1} mol cm^{-2}, and make the calculation at the electrode surface and at a distance 0.03 cm from the pulse source. (c) In the constant flux-induced diffusion, the time when the ion concentration at the electrode surface reduces to zero is called *transition time*, designated as τ. Derive an expression for τ and comment on its physical significance. Assume that in the constant flux experiment the concentration change is only caused by diffusion, i.e., the contribution of ion migration to concentration change is suppressed and therefore negligible. (Cf. Problem 4.15 in the textbook) (Xu)

Answer:

(a) In the case of constant flux-induced diffusion (cf. Eq. 4.72 in the textbook),

$$
c = c^{0} - \frac{\lambda}{\sqrt{D}} \left[2\sqrt{\frac{t}{\pi}} \exp\left(-\frac{x^{2}}{4Dt} \right) - \frac{x}{\sqrt{D}} erfc\left(\sqrt{\frac{x^{2}}{4Dt}} \right) \right] \tag{4.100}
$$

At the electrode surface, $x = 0$, and thus, $x^{2}/4Dt = 0$. Therefore, $\exp\left(x^{2}/4Dt \right) = 1$ and $\left(x/\sqrt{D} \right)erfc\sqrt{x^{2}/4Dt} = 0$. Under these conditions, Eq. (4.100) reduces to,

$$
c = c^{0} - 2\lambda\sqrt{\frac{t}{\pi D}} \tag{4.101}
$$

If the initial concentration value is $c^{0} = 1$ mol cm^{-3}, and $\lambda = 10^{-1}$ mol cm^{-2} s^{-1}, then,

$$
c = 1 \text{mol cm}^{-3} - 2\left(10^{-1} \text{ mol cm}^{-2} \text{s}^{-1} \right)\sqrt{\frac{t}{\pi\left(1.5 \times 10^{-5} \text{ cm}^{2}\text{s}^{-1} \right)}} \tag{4.102}
$$

The values of c at $\lambda = 10^{-1}$ and 1 mol cm^{-2} s^{-1} as a function of time are given in the next table, and Fig. 4.3 shows this variation:

t (µs)	Constant flux $\lambda = 10^{-1}$ mol cm^{-2} s^{-1} c (mol cm^{-3}) at $x = 0$	Constant flux $\lambda = 1$ mol cm^{-2} s^{-1} c (mol cm^{-3}) at $x = 0$
0	1.000	1.000
1	0.971	0.709
2	0.959	0.588
3	0.949	0.495
4	0.942	0.417
5	0.935	0.349
6	0.929	0.286

(b) In the case of instantaneous-pulse induced diffusion at a distance x from the pulse origin (cf. Eq. 4.91 in the textbook),

$$c = \frac{\lambda}{\sqrt{\pi D t}} \, exp\left(-\frac{x^2}{4 D t} \right) \tag{4.103}$$

At the electrode surface, $x = 0$, and thus, $\exp(x^2/4Dt) = 1$. Equation (4.103) reduces to,

$$c = \frac{\lambda}{\sqrt{\pi D t}} \tag{4.104}$$

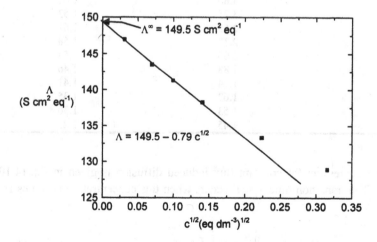

Figure 4.3. Concentration at the electrode surface as a function of time during constant flux.

In this case λ represents the *total concentration*, and thus has the units of mol cm^{-2}, in contrast with the constant-flux problem where it represented a *flux* given in mol cm^{-2} s^{-1}. Now, if $\lambda = 10^{-1}$ mol cm^{-2}, Eqs. (4.103) at $x = 0.03$ cm and Eq. (4.104) at $x = 0$ become respectively,

$$c = \frac{10^{-1} \text{ mol cm}^{-2}}{\sqrt{\pi\left(1.5 \times 10^{-5} \text{ cm}^2 \text{s}^{-1}\right)t}} \exp\left(-\frac{(0.03 \text{cm})^2}{4\left(1.5 \times 10^{-5} \text{ cm}^2 \text{s}^{-1}\right)t}\right) \quad (4.105)$$

and

$$c = \frac{10^{-1} \text{ mol cm}^{-2}}{\sqrt{\pi\left(1.5 \times 10^{-5} \text{ cm}^2 \text{s}^{-1}\right)t}} \quad (4.106)$$

The values of c at $\lambda = 10^{-1}$ mol cm^{-2} as a function of time are given in the next table, and Fig. 4.4 shows this variation:

t (s)	Pulse $\lambda = 10^{-1}$ mol cm^{-2} c (mol cm^{-3}) at $x = 0$ cm	Pulse $\lambda = 1$ mol cm^{-2} c (mol cm^{-3}) at $x = 0.03$ cm
5	6.51	0.32
10	4.60	1.03
20	3.25	1.53
30	2.65	1.61
40	2.30	1.58
50	2.06	1.53
60	1.88	1.46
70	1.74	1.41
80	1.62	1.35
90	1.53	1.30
100	1.45	1.25

(c) The expression for constant-flux induced diffusion is given in Eq. (4.101) above. The transition time, $t = \tau$, occurs when this concentration becomes zero, i.e.,

$$c = c^0 - 2\lambda\sqrt{\frac{\tau}{\pi D}} = 0 \quad (4.107)$$

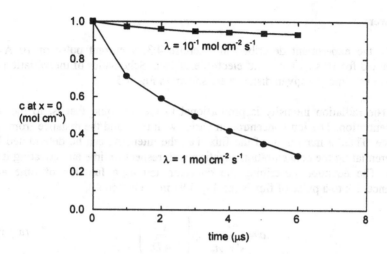

Figure 4.4. Concentration at the electrode surface as a function of time during an instantaneous pulse.

The transition time is, then,

$$\tau = \pi D \left(\frac{c^0}{2\lambda} \right)^2 \qquad (4.108)$$

Comment: From these two expressions it can be seen that as $t > \tau$, concentration assumes negative values, $c < 0$, which is only the mathematical solution of the equation but has no physical significance. Therefore, constant flux problems are time-restricted processes. They do not exist beyond τ.

4.31 In the same experiment with radioactive material described in Exercise 4.3, it was found that at a certain time the Geiger counter registered a maximum ion flux, i.e., the intensity of the radiation had a maximum with respect to time. It was also found that by placing the Geiger counter farther away from the electrode, the time at which the maximum occurred became longer and the peak intensity rapidly decreased. (a) Draw diagrams of *current* vs. *time* plots at $x = 1$ cm, $x > 1$ cm, and $x \gg 1$ cm, where x is the distance of the Geiger counter from the electrode. (b) Justify the observation presented here. (c) Evaluate its usefulness in experimentally measuring diffusion coefficients of ions. (Cf. Problem 4.1 in the textbook) (Xu)

Answer:

(a) In the experiment described in Exercise 4.3, a current pulse of 10 A is generated for 10 s on a 0.1 cm² electrode surface. Schematics of the variation of current vs. time at a given distance are shown in Fig. 4.5.

(b) The radiation intensity is proportional to the ion flux, that is, to the ion concentration. The ion concentration varies with time and the distance from the source. Thus, a maximum of the flux, i.e., the intensity, can be determined by differentiating the concentration equation with respect to time and equating it to zero. The equation describing the concentration as a function of time and distance due to a pulse of flux is (cf. Eq. 4.91 in the textbook),

$$c = \frac{\lambda}{\sqrt{\pi Dt}} \exp\left(-\frac{x^2}{4Dt} \right)$$
(4.109)

and its derivative,

$$\frac{\partial c}{\partial t} = \frac{\lambda}{\sqrt{\pi Dt}} \frac{\partial}{\partial t} \exp\left(-\frac{x^2}{4Dt} \right) + \exp\left(-\frac{x^2}{4Dt} \right) \frac{\partial}{\partial t}\left(\frac{\lambda}{\sqrt{\pi Dt}} \right)$$

$$= \left(\frac{1}{4} \frac{\lambda x^2 t^{-5/2}}{\sqrt{\pi D^3}} - \frac{1}{2} \frac{\lambda t^{-3/2}}{\sqrt{\pi D}} \right) \exp\left(-\frac{x^2}{4Dt} \right)$$
(4.110)

Equating this derivative to zero,

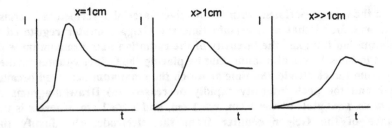

Figure 4.5. Variation of current with time at different distances from the electrode according to the experimental parameters described in Problem 4.31.

$$\left(\frac{1}{4} \frac{\lambda x^2 t^{-5/2}}{\sqrt{\pi D}^3} - \frac{1}{2} \frac{\lambda t^{-3/2}}{\sqrt{\pi D}} \right) exp \left(-\frac{x^2}{4Dt} \right) = 0 \qquad (4.111)$$

or

$$\frac{x^2}{2D} \frac{1}{t} - 1 = 0 \qquad (4.112)$$

Solving for t,

$$t_{max} = \frac{x^2}{2D} \qquad (4.113)$$

which is the "rush" hour for the ion flux at x. As x increases (Geiger moved away form the electrode) the time corresponding to this ionic "peak traffic" is delayed. An interesting fact is that this rush hour is actually determined exactly by the Einstein-Smoluchowski equation. This should not be a surprise as it has been proved that Einstein-Smoluchowski equation describes the diffusion behavior of the majority of the ions, and this majority quantity of ions constitute the "peak traffic". Now, inserting Eq. (4.113) into Eq. (4.109), gives the maximum ion concentration, i.e.,

$$c_{max} = \frac{\lambda}{\sqrt{\pi/2}} \frac{1}{x} exp \left(-\frac{1}{2} \right) = 0.484 \frac{\lambda}{x} \qquad (4.114)$$

As the Geiger moves away from the electrode, the intensity of radiation, which is proportional to the ion flux, decreases rapidly according to Eq. (4.114).

(c) The above experiment could be used to measure the diffusion coefficients of ions, but the ions do not have to be radioactive as long as there are other ways to determine the change of ion concentration (visible-UV, refractive indices, etc.)

4.32 The unit flux, i.e., 1 mol cm^{-2} s^{-1}, has been used in an attempt to simplify the solution of the partial-derivative equation of Fick's second law. (a) Calculate, for an univalent ion, the current density and the electrical field this flux will cause. Make any reasonable assumption for your calculations. Is this current density achievable in a real experiment? (b) 1 mA cm^{-2} is the current density normally used in electrochemistry laboratories. What is the flux in this case? (Cf. Exercise 4.1 in the textbook) (Xu)

Answer:

(a) The unit flux means that 1 mol of ions passes through an area of 1 cm^2 in 1 s. Because each mole of uni-valent ions carries 1 Faraday or 96500 C of charge, then the charge flux is:

$$\vec{j} = \left(1 \text{mol cm}^{-2} \text{s}^{-1} \right) \left(96500 \text{C mol}^{-1} \right) \left(\frac{1 \text{A}}{\text{C s}^{-1}} \right) = 96500 \text{A cm}^{-2} \quad (4.115)$$

Usually the conductivity of the electrochemical systems under study is below 10^{-1} S cm^{-1}. If the linear relation between field and flux holds (most unlikely!), then the field $\vec{X}$ will be given as,

$$\vec{X} = \frac{\vec{j}}{\sigma} = \frac{9.65 \times 10^4 \text{ A cm}^{-2}}{10^{-1} \Omega^{-1} \text{cm}^{-1}} = 9.65 \times 10^4 \text{ A cm}^{-1} \Omega \times \frac{1 \text{V}}{1 \text{A} \Omega} \quad (4.116)$$
$$= 9.65 \times 10^4 \text{ V cm}^{-1}$$

Almost no electrolyte system can withstand this high field. Therefore, unit flux is only a thought convenience and not feasible in experiments.

(b) When $\vec{j} = 0.001$ A cm^{-2}, then (cf. Eq. 4.127 in the textbook),

$$\vec{J} = \frac{\vec{j}}{zF} = \frac{0.001 \text{A cm}^{-2}}{|1| 96500 \text{C mol}^{-1}} \times \frac{1 \text{C s}^{-1}}{1 \text{A}} = 1.036 \times 10^{-8} \text{ mol cm}^{-2} \text{s}^{-1} \quad (4.117)$$

4.33 A saturated solution of silver chloride has a resistance of 67.953 kΩ when placed in a conductance cell of constant 0.180 cm^{-1} at 25 ^{0}C. The resistance of the water used as solvent is 212.180 kΩ measured in the same cell. Calculate the solubility, S, of the salt at 25 ^{0}C. Consider the salt completely dissociated in its saturated solution in water. The equivalent conductivity at infinite dilution of Ag$^+$ and Cl$^-$ are 61.92 and 76.34 S cm^2 eq^{-1}, respectively. (Cf. Exercise 4.37 in the textbook) (Constantinescu)

Answer:

The solubility can be determined from conductance measurements. The equivalent conductivity is given by (cf. Table 4.55 in the textbook),

$$\Lambda = \frac{\sigma}{cz} \tag{4.118}$$

If S is the solubility of the salt AgCl, then, $S_{AgCl} = c_{AgCl}$, and

$$S_{AgCl} = \frac{\sigma_{AgCl}}{\Lambda_{AgCl} z} \tag{4.119}$$

where σ_{AgCl} is the specific conductivity of the salt. This parameter, σ_{AgCl}, can be determined from the specific conductivity of the AgCl solution, which consists of two contributions,

$$\sigma_{soln} = \sigma_{AgCl} + \sigma_{water} \tag{4.120}$$

The specific conductivity of the saturated solution is given by (cf. Fig. 4.55 in the textbook)

$$\sigma_{soln} = \frac{1}{R}\frac{l}{A} = \frac{k}{R} = \frac{0.180\,cm^{-1}}{67953\,\Omega} = 2.65 \times 10^{-6}\ S\,cm^{-1} \tag{4.121}$$

and that of water,

$$\sigma_{water} = \frac{0.180\,cm^{-1}}{212180\,\Omega} = 0.848 \times 10^{-6}\ S\,cm^{-1} \tag{4.122}$$

Therefore, the specific conductivity due to the salt is, from Eq. (4.120),

$$\sigma_{AgCl} = \sigma_{soln} - \sigma_{water} = 2.65 \times 10^{-6} - 0.848 \times 10^{-6} = 1.80 \times 10^{-6}\ S\,cm^{-1} \tag{4.123}$$

On the other hand, the equivalent conductivity of the salt is given by

$$\Lambda_{AgCl} \approx \Lambda^0_{AgCl} = \lambda^0_{Ag^+} + \lambda^0_{Cl^-} = 61.92 + 76.3 = 138.26\ S\,cm^2\,eq^{-1} \tag{4.124}$$

Therefore, from Eq. (4.119)

$$S_{AgCl} = \frac{1.80 \times 10^{-6} \, \text{S cm}^{-1}}{|1|138.26 \, \text{S cm}^2 \, \text{eq}^{-1}} \times \frac{1000 \, \text{cm}^3}{1 \, \text{dm}^3} = 1.302 \times 10^{-5} \, \text{eq dm}^{-3} \quad (4.125)$$

Comment: The method described here gives the ionic concentration in the saturated solution, and it is only when dissociation is virtually complete that the result is identical with the solubility. The solution has to be sufficiently diluted (i.e., the solubility has to be small) for the equivalent conductivity to be similar to the value at infinite dilution.

4.34 Calculate the relaxation times of the ionic atmospheres of 0.1 N solutions of a 1:1 electrolyte in nitrobenzene, ethyl alcohol, and ethylene dichloride. Consider the Debye and Falkenhagen equation, which indicates that the relaxation time, τ_R, is related to the frictional coefficients K_+ and K_- of the two ions constituting a binary electrolyte:

$$\tau_R (s) = \frac{2 K_+ K_-}{K_+ + K_-} \frac{\left(\kappa^{-1} \right)^2}{kT} \qquad (4.126)$$

where

$$K_i = e_0 zF / \lambda, \qquad (4.127)$$

k is the Boltzmann constant, and κ^{-1} is the thickness of the ionic atmosphere. (Constantinescu)

Answer:

Equation (4.126) can be written as:

$$\tau_R (s) = \frac{2}{\dfrac{1}{K_+} + \dfrac{1}{K_-}} \frac{\left(\kappa^{-1} \right)^2}{kT} \qquad (4.128)$$

Substituting K_+ and K_- from Eq. (4.127), into Eq. (4.128)

$$\tau_R\left(s\right)=\frac{2e_0zF}{\lambda_+ + \lambda_-}\frac{\left(\kappa^{-1}\right)^2}{kT}=\frac{2e_0zF}{\Lambda}\frac{\left(\kappa^{-1}\right)^2}{kT}$$

(4.129)

$$=3.09 \times 10^{-14}\frac{z\left(\kappa^{-1}\right)^2}{\Lambda kT}$$

where κ^{-1} is in cm, Λ in S cm^2 eq^{-1}, k in J K^{-1} and T in K. For most solutions other than acids and bases, Λ is about 120 S cm^2 eq^{-1} at 25 ^{0}C (cf. Table 4.10 in the textbook). The thickness of the ionic atmosphere, κ^{-1}, is given by Eq. (3.43) in the textbook. For a 0.1 N 1:1 electrolyte at 25 ^{0}C, κ^{-1} in meters is given by (cf. Exercise 3.6)

$$\kappa^{-1}=1.086 \times 10^{-10}\sqrt{\varepsilon}$$

(4.130)

For nitrobenzene with an $\varepsilon_{nitrobenzene}$ = 34.8, κ^{-1} = 6.40x10^{-8} cm (cf. Exercise 3.6),

$$\tau_{R\,nitrobenzene}=30.8 \times 10^{-8}\frac{(1)\left(6.40 \times 10^{-8}\,cm\right)^2}{120\,S\,cm^2\,eq^{-1}\left(1.381 \times 10^{-16}\,erg\,K^{-1}\right)(298K)}$$

$$=2.55 \times 10^{-10}\,s$$

(4.131)

In the same way for the other solvents,

	ε	κ^{-1}(cm)	τ_R (s)
Nitrobenzene	34.8	6.40 x10^{-8}	2.55 x10^{-10}
Ethyl alcohol	24.3	5.352x10^{-8}	1.786x10^{-10}
Ethylene dichloride	10.4	3.501x10^{-8}	0.764x10^{-10}

Comment: The existence of a finite time of relaxation means that the ionic atmosphere surrounding a moving ion is not a symmetrical one. The charge density behind the ion is greater than that in front. The asymmetry of the ionic cloud due to the time of relaxation results in a retardation of the ion moving under the influence of an applied electric field. The influence on the speed of an ion is called *relaxation effect*. The P. Debye and H. Falkenhagen equation can be found in Physik. Z. **29** (1928)121,401; H Flakenhagen and J. W. Williams, Z. Physik. Chem. **137** (1928)399; and J. Phys. Chem., **33** (1929)1121.

4.35 The values of molar conductivity at infinite dilution for HCl, NaCl and sodium acetate (NaAc) are 420, 126, and 91 S cm^2 mol^{-1}, respectively. The resistance of a conductivity cell filled with 0.1 M acetic acid (HAc) is 520 Ω (Solution 1), and drops to 122 Ω when enough NaCl is added to make the solution 0.1 M NaCl (Solution 2). Calculate (a) the cell constant and (b) the concentration of hydrogen ion (pH) in Solution 1. (Contractor)

Answer:

(a) The cell constant is given by the ratio $k = l / A$. From the specific conductivity equation (cf. Eq. 4.134 in the textbook),

$$\sigma = \frac{1}{R}\frac{l}{A} = Gk \qquad (4.132)$$

Therefore, for NaCl,

$$\sigma_{NaCl} = G_{NaCl} k \qquad (4.133)$$

where G_{NaCl} is the conductance of NaCl. Now, the total conductance, G, in a cell is given by the addition of the conductances of its components. In the case of the 0.1 M NaCl + 0.1 M HAc solution, the total conductance is $G = 1/R_2$, and also, $G = G_{NaCl} + G_{HAc}$. Therefore,

$$G_{NaCl} = \frac{1}{122\,\Omega} - \frac{1}{520\,\Omega} = 0.00627\,\Omega^{-1} \qquad (4.134)$$

The specific conductivity of NaCl can be approximated as,

$$\sigma_{NaCl} = \Lambda_{m,NaCl}^{\infty} c_{NaCl} \qquad (4.135)$$

$$= \left(126\,\text{S cm}^2\,\text{mol}^{-1} \right)\left(0.1\,\text{mol dm}^{-3} \right) \times \times \frac{1\,\text{dm}^3}{1000\,\text{cm}^3} = 0.0126\,\text{S cm}^{-1}$$

Substituting the corresponding values of G_{NaCl} and σ_{NaCl} from Eqs. (4.134) and (4.135) into Eq. (4.133) and solving for k,

$$k = \frac{\sigma_{NaCl}}{G_{NaCl}} = \frac{0.0126\,\text{S cm}^{-1}}{0.00627\,\text{S}} = 2.01\text{cm}^{-1} \tag{4.136}$$

(b) From the equilibrium reaction for acetic acid,

$$HAc \underset{\leftarrow}{\overset{\rightarrow}{}} H^+ + Ac^- \tag{4.137}$$

the concentrations of H^+ and Ac^- are given by

$$c_{H^+} = c_{Ac^-} = \alpha\, c_{HAc} \tag{4.138}$$

where α is the degree of dissociation of 0.1 M HAc, and is given by,

$$\alpha = \frac{\Lambda_m}{\Lambda_m^\infty} \tag{4.139}$$

The molar conductivity of 0.1 M HAc solution, $\Lambda_{m,1}$, is given by (cf. Eq. 4.136 in the textbook) $\Lambda_{m,1} = \sigma_1/c_1$ and the specific conductivity, σ_1, by (cf. Eq. 4.134 in the textbook), $\sigma_1 = G_1 k$. Therefore,

$$\Lambda_{m,1} = \frac{kG_1}{c_1} = \frac{2.01\text{cm}^{-1}}{(520\,\Omega)\left(0.1\,\text{mol dm}^{-3}\right)} \times \frac{1000\,\text{cm}^3}{1\,\text{dm}^3} \tag{4.140}$$

$$= 38.7\,\text{S cm}^2\,\text{mol}^{-1}$$

The molar conductivity at infinite dilution can be determined by Kohlrausch's law of the independent migration of ions, (cf. Sec. 4.3.10 in the textbook) i.e.,

$$\Lambda_{m,HAc}^\infty = \lambda_{m,H^+}^\infty + \lambda_{m,Ac^-}^\infty = \Lambda_{m,HCl}^\infty + \Lambda_{m,NaAc}^\infty - \Lambda_{m,NaCl}^\infty \tag{4.141}$$

$$= (420 + 91 - 126)\,\text{S cm}^2\,\text{mol}^{-1} = 385\,\text{S cm}^2\,\text{mol}^{-1}$$

Substituting Eqs. (4.140) and (4.141) into Eq. (4.139) gives the degree of dissociation of acetic acid in Solution 1,

$$\alpha = \frac{38.7 \, \text{S cm}^2 \, \text{mol}^{-1}}{385 \, \text{S cm}^2 \, \text{mol}^{-1}} = 0.100 \tag{4.142}$$

and, thus, the concentration of H^+ is, from Eq. (4.138),

$$c_{H^+} = (0.100)(0.1 \, \text{M}) = \textbf{0.01 M} \tag{4.143}$$

or

$$pH_1 = -\log c_{H^+} \, (\text{M}) = -\log(0.01 \, \text{M}) = \textbf{2} \tag{4.144}$$

4.36 The equivalent conductivity at infinite dilution of the divalent copper ion is 55 S cm^2 eq^{-1}, and its ionic radius, 72 pm. Calculate the primary solvation number of Cu^{+2}. The radius of a water molecule is 138 pm and the viscosity of the solution is 10^{-2} poise. (Cf. Exercise 4.19 in the textbook) (Herbert)

Answer:

The expression of the primary hydration number as the ratio between the volume of the hydration sheath and the volume of the water molecule is (cf. Eq. 2.23 in the textbook),

$$n_h = \frac{r_S^3 - r_i^3}{r_w^3} \tag{4.145}$$

where r_S is the Stoke's radius, i.e., the radius of the ion and its primary solvation shell, r_i is the crystallographic radius of the ion, and r_w is the radius of the water molecule. The Stoke's radius can be obtained from the Einstein-Stokes-Nernst equation (cf. Eq. 4.196 in the textbook), i.e., applied to a single ion,

$$\eta \lambda_i = \frac{z_i e_0 F}{6 \pi r_S} \tag{4.146}$$

Therefore (cf. Exercise 4.16),

$$r_S = \frac{z_i e_0 F}{6\pi\eta\lambda_i} = \frac{|2|\left(1.602 \times 10^{-19}\, C\right)\left(96500\, C\, mol^{-1}\right)}{6\pi\left(10^{-2}\, poise\right)\left(55\, S\, cm^2\, eq^{-1}\right)} \tag{4.147}$$

$$\times \frac{1\, poise}{10^{-7}\, \Omega\, C^2\, cm^{-3}} = 2.98 \times 10^{-8}\, cm$$

Substituting this value as well as the values of r_i and r_w into Eq. (4.145) gives

$$n_h = \frac{\left(2.98 \times 10^{-8}\, cm\right)^3 - \left(0.72 \times 10^{-8}\, cm\right)^3}{\left(1.38 \times 10^{-8}\, cm\right)^3} = 9.93 \tag{4.148}$$

Comment: The mobility method is one of the most adequate for the determination of the primary solvation number of ions, but the geometric simplification may lead to some error.

4.37 A definition for specific conductivity frequently cited in electrochemical literature (cf. Eq. 4.160 in the textbook) is,

$$\sigma = \sum_i n_i \left(u_{conv}\right)_i z_i e_0 \tag{4.149}$$

where n_i is the number of ions in a unit volume, $(u_{conv})_i$ is the conventional mobility, z_i is the valence state, and e_0 is the elemental charge. Starting from the definition of specific activity, i.e.,

$$\sigma = \frac{\vec{j}}{\vec{X}} \tag{4.150}$$

derive Eq. (4.149). (Cf. Exercise 4.14 in the textbook) (Xu)

Answer:

The specific conductivity of a single ion is given by (cf. Eq. 4.128 in the textbook) is,

$$\sigma_i = \frac{\vec{j}_i}{\vec{X}} \tag{4.151}$$

where $\vec{j}_i$ is the current density caused by the transport of the ionic species i, and $\vec{X}$ is the electric field. The current density is defined as $\vec{j}_i = \vec{I}_i / A$, where A is the area of the ion-flux cross. Furthermore, the current is defined as the quantity of charge, Q_i, crossing the area A in the unit time t, i.e., $\vec{I}_i = Q_i / t$. Thus,

$$\vec{j}_i = \frac{Q_i}{tA} \tag{4.152}$$

Substituting $\vec{j}_i$ into Eq. (4.151), gives

$$\sigma_i = \frac{Q_i}{tA\vec{X}} \tag{4.153}$$

Further, the charge Q_i is given by

$$Q_i = (number\ of\ ions\ crossing\ the\ area\ A) \times (charge\ of\ the\ ion) \tag{4.154}$$

where

$$(number\ of\ ions\ crossing\ the\ area\ A)$$
$$= (volume) \times (number\ of\ ions\ per\ unit\ volume) = Vn_i \tag{4.155}$$

The total volume needed in Eq. (4.155) is given by (see Fig. 4.6)

$$V = Av_d t \tag{4.156}$$

Therefore, the charge Q_i becomes,

$$Q_i = Av_{d,i} t n_i z_i e_0 \tag{4.157}$$

Substituting Q_i into Eq. (4.153),

$$\sigma_i = \frac{v_{d,i} n_i z_i e_0}{\vec{X}} \tag{4.158}$$

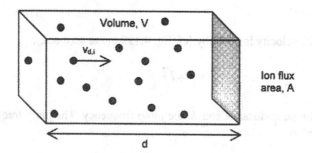

Figure 4.6. Cross area of an ionic flux.

Now, the conventional mobility is defined as (cf. Sec. 4.4.3 in the textbook)

$$\left(u_{conv}\right)_i = \frac{v_{d,i}}{\vec{X}} \qquad (4.159)$$

Therefore, Eq. (4.158) becomes

$$\sigma_i = \left(u_{conv}\right)_i n_i z_i e_0 \qquad (4.160)$$

Applying Kohlrausch's independent law, the total conductivity due to the movement of all the ionic species is (cf. Eq. 4.143 in the textbook),

$$\sigma = \sum_i \sigma_i = \sum_i \left(u_{conv}\right)_i n_i z_i e_0 \qquad (4.161)$$

which is the definition of specific conductivity found in literature.

4.38 Using the data given in the table below, calculate the activation energy for diffusion for the electrolytes KCl and tetraethylammonium picrate and comment on their relative magnitudes. (Contractor)

Temperature (^{0}C)	Λ^0 of KCl (S cm^2 eq^{-1})	Λ^0 of Tetraethylammonium-picrate (S cm^2 eq^{-1})
0	81.8	31.2
18	129.8	53.2
100	406.0	196.5

Answer:

The drift velocity is given by (cf. Eq. 4.199 in the textbook),

$$v_d = l\vec{k} \tag{4.162}$$

where l is the jump distance and $\vec{k}$ the jump frequency. The jump frequency can be expressed as

$$\vec{k} = Ae^{-E_a/RT} \tag{4.163}$$

where E_a is the activation energy and A the pre-exponential factor, independent of temperature. Now, the velocity is equal to the conventional mobility (cf. Eq. 4.151 in the textbook),

$$v_d = u_{conv}, \tag{4.164}$$

and u_{conv} is related to the equivalent conductivity by (cf. Eq. 4.163 in the textbook)

$$\Lambda = F\sum_i u_{conv,i} = Fu_{conv} \tag{4.165}$$

or at infinite dilution,

$$\Lambda^0 = Fu_{conv} \tag{4.166}$$

Therefore, putting Eqs. (4.162)-(4.164) and (4.166) together,

$$\Lambda^0 = FlAe^{-E_a/RT} \quad \text{or} \quad ln\,\Lambda^0 = ln\,FlA - \frac{E_a}{R}\frac{1}{T} \tag{4.167}$$

According to Eq. (4.167), a graph of $log\Lambda^0$ vs. $1/T$ should give a straight line with slope E_a/R. The values of $log\Lambda^0$ at different temperatures for the two electrolytes are given below

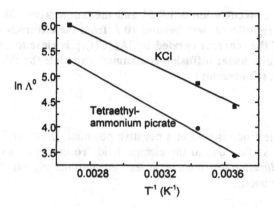

Figure 4.7. Graph of ln Λ^0 vs. T^{-1} for two electrolytes for the determination of their diffusion-activation energy.

$1/T$ (K^{-1})	$\log \Lambda^0$ (KCl)	$\log \Lambda^0$ (Tetraethyl ammonium picrate)
0.00366	4.404	3.440
0.00344	4.866	3.974
0.00268	6.006	5.281

The corresponding plots and parameters are given in Fig. 4.7 and in the following table:

	Slope	E_a (kJ K^{-1} mol^{-1})
KCl	1599	13.29
Tetraethylammonium picrate	1837	15.28

Comment: A higher value of activation energy is reasonable for the tetraethylammonium picrate because of its much larger molecular size.

4.39 (a) Calculate the minimum concentration gradient necessary to obtain a flux of [Ag(CN)₂]⁻ *towards* a silver electrode when the electrode is held at –0.5 V in a point 10 Å from the electrode surface. The concentration of [Ag(CN)₂]⁻ at the surface of the electrode can be assumed to be 0.01 M,

while the bulk concentration is 1.0 M and the temperature 300 K. Assume that there is a significant field beyond 10 Å from the electrode surface, and the fraction of the current carried by $[Ag(CN)_2]^-$ is close to one (Fig. 4.8). (b) If semi-infinite linear diffusion is assumed, calculate the thickness of the diffusion layer. (Contractor)

Answer:

(a) Since the electrode is held at a negative potential, $[Ag(CN)_2]^-$ will migrate away from the surface due to the electric field. To obtain a flux towards the electrode, the *diffusion* flux must be at least equal to the *migration* flux (cf. Eq. 4.226 in the textbook),

$$D\frac{dc}{dx} = \frac{Dc_0}{RT}zF\vec{X} \tag{4.168}$$

where c_0 is the surface concentration and $\vec{X}$ the electric field, i.e., $\vec{X} = 0.5\,\text{V}/10\times10^{-8}\,\text{cm}$. Thus,

$$\frac{dc}{dx} = \frac{\left(0.01\,\text{mol dm}^{-3}\right)|1|\left(96500\,\text{C mol}^{-1}\right)\left(0.05\times10^{8}\,\text{V cm}^{-1}\right)}{\left(8.314\,\text{J K}^{-1}\text{mol}^{-1}\right)\left(300\,\text{K}\right)}\times\frac{1\text{J}}{1\text{CV}}$$

$$= 1.9\times10^{6}\,\text{M cm}^{-1} \tag{4.169}$$

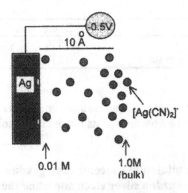

Figure 4.8. Schematic of the process described in Problem 4.39

(b) Assuming semi-infinite linear diffusion,

$$\frac{dc}{dx} = \frac{c_b - c_o}{\delta} \tag{4.170}$$

where c_b is the bulk concentration, c_o the concentration at the surface of the electrode, and δ the diffusion layer thickness. Then,

$$\delta = \frac{c_b - c_o}{dc/dx} = \frac{1.0\,M - 0.01\,M}{1.9 \times 10^6 \, M\,cm^{-1}} = 5.1 \times 10^{-7} \, cm \tag{4.171}$$

4.40 Estimate the diffusion coefficient of Na^+ and Cl^- in water at 298 K from the equivalent conductivity at infinite dilution of NaCl, Λ^0_{NaCl} = 126.46 S cm² eq⁻¹ and the cation transport number, t^0_{Na+} = 0.396. (Cf. Exercise 4.7) (Herbert)

Answer:

From the Einstein relation (cf. Eq. 4.172 in the textbook) the diffusion coefficient is related to the absolute mobility by,

$$D_i = \bar{u}_{abs,i} \, kT \tag{4.172}$$

The absolute mobility is related to the conventional mobility by (cf. Eq. 4.152 in the textbook),

$$\bar{u}_{abs,i} = \frac{u_{conv,i}}{z_i e_0} \tag{4.173}$$

which is given in terms of the equivalent conductivity by (cf. Eq. 4.163 in the textbook),

$$u_{conv,i} = \frac{\lambda_i}{F} \tag{4.174}$$

Now, the transport number of the species i is given by (cf. Eq. 4.236 in the textbook),

$$t_i = \frac{u_i}{\sum\limits_i u_i} \tag{4.175}$$

Substituting Eq. (4.174) into Eq. (4.175),

$$t_i = \frac{\lambda_i}{\sum\limits_i \lambda_i} = \frac{\lambda_i}{\Lambda} = \frac{Fu_{conv,i}}{\Lambda} \tag{4.176}$$

Fom Eqs. (4.176) and (4.173), Eq. (4.172) becomes,

$$D_i = \frac{u_{conv,i}kT}{z_i e_0} = \frac{t_i \Lambda^0 kT}{Fz_i e_0} \tag{4.177}$$

For Na^+,

$$D_{Na^+} = \frac{t_{Na^+} \Lambda^0 kT}{Fz_{Na^+} e_0} = \frac{(0.396)\left(126.46\,S\,cm^2\,eq^{-1}\right)\left(1.381 \times 10^{-23}\,J\,K^{-1}\right)}{\left(96500\,C\,eq^{-1}\right)(1)}$$

$$\times \frac{(298\,K)}{\left(1.602 \times 10^{-19}\,C\right)} \times \frac{1C}{1Vs} \times \frac{1CV}{1J} = 1.33 \times 10^{-5}\,cm^2\,s^{-1} \tag{4.178}$$

Similarly, for the Cl^- ion, considering that $t_{Cl^-} = 1 - t_{Na^+} = 1 - 0.396$, gives, $D_{Cl^-} = 2.03 \times 10^{-5}\,cm^2\,s^{-1}$.

Another way to calculate the diffusion coefficient is through the Nernst-Einstein equation after one has calculated the equivalent conductivities of the ions.

Comment: The answer is sustained by the assumption that chemical and electrical forces encounter the same resistance in moving a given ion through the solution, even if the mechanistic picture of a force acting on a single particle cannot be applied to diffusion.

4.41 A student has to determine the equivalent conductivity at infinite dilution of KCl, NaCl, KNO₃, and NaNO₃ solutions, as well as the transport

numbers of the ions in these solutions. He managed to determine only $\Lambda^0(KNO_3)$, $\Lambda^0(NaNO_3)$, $t^0(Na^+/NaCl)$, and $t^0(K^+/KCl)$, and wrote them in a table:

	NaNO$_3$	KNO$_3$	NaCl	KCl
Λ^0 (S cm^2 eq^{-1})	121.4	144.9	-	-
t_+	-	-	0.396	0.490
t_-	-	-	-	-

Assuming that the determined values are correct, help him fill in the spaces in the table without doing any further experiment. (Cf. Exercise 4.25) (Herbert)

Answer:

Kohlrausch's law of the independent migration of ions leads to (cf. Eq. 4.3.10 in the textbook),

$$\lambda^0_{K^+} - \lambda^0_{Na^+} = \Lambda^0_{KCl} - \Lambda^0_{NaCl} = \Lambda^0_{KNO_3} - \Lambda^0_{NaNO_3} \qquad (4.179)$$

Also (cf. Eq. 4.175):

$$\lambda^0_{K^+} = t^0_{K^+(KCl)}\Lambda^0_{KCl} \quad \text{and} \quad \lambda^0_{Na^+} = t^0_{Na^+(NaCl)}\Lambda^0_{NaCl} \qquad (4.180)$$

Equation (4.179) can also be written as,

$$\lambda^0_{K^+} - \lambda^0_{Na^+} = \Lambda^0_{KNO_3} - \Lambda^0_{NaNO_3}$$
$$= t^0_{K^+(KCl)}\Lambda^0_{KCl} - t^0_{Na^+(NaCl)} - \Lambda^0_{NaCl} \qquad (4.181)$$

Equations (4.179) and (4.180) form a set of equations with variables Λ^0_{NaCl} and Λ^0_{KCl}. Thus, multiplying Eq. (4.179) by $- t^0_{Na^+(NaCl)}$ and adding the two equations,

$$-t^0_{Na^+(NaCl)}\Lambda^0_{KCl} + t^0_{Na^+(NaCl)}\Lambda^0_{NaCl} = -t^0_{Na^+(NaCl)}\Lambda^0_{KNO_3} + t^0_{Na^+(NaCl)}\Lambda^0_{NaNO_3}$$
$$(4.182a)$$

$$t^0_{K^+(KCl)}\Lambda^0_{KCl} - t^0_{Na^+(NaCl)}\Lambda^0_{NaCl} = \Lambda^0_{KNO_3} - \Lambda^0_{NaNO_3} \qquad (4.182b)$$

$$\left(t^0_{K^+(KCl)} - t^0_{Na^+(NaCl)}\right)\Lambda^0_{KCl} = \left(1 - t^0_{Na^+(NaCl)}\right)\left(\Lambda^0_{KNO_3} - \Lambda^0_{NaNO_3}\right)$$
$$(4.183)$$

or

$$\Lambda^0_{KCl} = \frac{\left(1 - t^0_{Na^+(NaCl)}\right)\left(\Lambda^0_{KNO_3} - \Lambda^0_{NaNO_3}\right)}{t^0_{K^+(KCl)} - t^0_{Na^+(NaCl)}} \qquad (4.184)$$

$$= \frac{(1 - 0.396)(144.9 - 121.4)\,\text{S cm}^2\,\text{eq}^{-1}}{0.490 - 0.396} = 151.0\,\text{S cm}^2\,\text{eq}^{-1}$$

and from Eq. (4.179),

$$\Lambda^0_{NaCl} = \Lambda^0_{KCl} - \Lambda^0_{KNO_3} - \Lambda^0_{NaNO_3} \qquad (4.185)$$

$$= 151.0 - 144.9 + 121.4 = 127.5\,\text{S cm}^2\,\text{eq}^{-1}$$

From equations similar to those in Eq. (4.180), the transport numbers of the other ions are,

$$t^0_{Na^+(NaNO_3)} = \frac{\lambda^0_{Na^+}}{\Lambda^0_{NaNO_3}} = \frac{t^0_{Na^+(NaCl)}\Lambda^0_{NaCl}}{\Lambda^0_{NaNO_3}} \qquad (4.186)$$

$$= \frac{(0.396)\left(127.5\,\text{S cm}^2\,\text{eq}^{-1}\right)}{121.4\,\text{S cm}^2\,\text{eq}^{-1}} = 0.416$$

$$t^0_{K^+(KNO_3)} = \frac{\lambda^0_{K^+}}{\Lambda^0_{KNO_3}} = \frac{t^0_{K^+(KCl)}\,\Lambda^0_{KCl}}{\Lambda^0_{KNO_3}}$$

(4.187)

$$= \frac{(0.490)\left(151.0\,\text{S cm}^2\,\text{eq}^{-1}\right)}{144.9\,\text{S cm}^2\,\text{eq}^{-1}} = 0.511$$

The anion transport numbers are computed by subtracting the cation transport number from unity and the resulting values are:

	NaNO$_3$	KNO$_3$	NaCl	KCl
Λ^0 (S cm^2 eq^{-1})	121.4	144.9	127.5	151.0
t^0_+	0.416	0.511	0.396	0.490
t^0_-	0.584	0.489	0.604	0.510

Comment: At higher concentrations, ion-ion interactions make λ differ from λ^0, and the law of independent migration of ions is not strictly valid anymore.

4.42 The transport number of Ca^{+2} in a CaCl$_2$ solution is $t_{Ca^{+2}/CaCl_2} = 0.438$, and that of K$^+$ in a KCl solution is $t_{K^+/KCl} = 0.490$. Calculate the transport number of Ca^{+2} in a solution containing both, 0.001 M CaCl$_2$ and 0.01 M KCl. Neglect the variation of t and Λ with concentration. (Cf. Exercise 4.26 in the textbook) (Herbert)

Answer:

The transport number is related to the equivalent conductivity by the relation (cf. Eq. 4.163 and Section 4.5.2 in the textbook)

$$t_i = \frac{z_i c_i \lambda_i}{\sum\limits_i z_i c_i \lambda_i}$$

(4.188)

Therefore, the transport number of Ca^{+2} in the mixture of solutions (0.001 M CaCl$_2$ + 0.01 M KCl) is,

$$t_{Ca^{+2}/mixture} = \frac{z_{Ca^{+2}} c_{Ca^{+2}} \lambda_{Ca^{+2}}}{z_{Ca^{+2}} c_{Ca^{+2}} \lambda_{Ca^{+2}} + z_{K^+} c_{K^+} \lambda_{K^+} + z_{Cl^-} c_{Cl^-} \lambda_{Cl^-}} \tag{4.189}$$

The next step is to calculate the different λ_i in Eq. (4.189). For each separate solution, the corresponding transport numbers are given by,

$$t_{Ca^{2+}/CaCl_2} = \frac{\lambda_{Ca^{+2}}}{\Lambda_{CaCl_2}} \tag{4.190}$$

$$t_{Cl^-/CaCl_2} = \frac{\lambda_{Cl^-}}{\Lambda_{CaCl_2}} \tag{4.191}$$

$$t_{K^+/KCl} = \frac{\lambda_{K^+}}{\Lambda_{KCl}} \tag{4.192}$$

$$t_{Cl^-/KCl} = \frac{\lambda_{Cl^-}}{\Lambda_{KCl}} \tag{4.193}$$

Combining Eqs. (4.188) and (4.189),

$$\frac{\lambda_{Ca^{+2}}}{\lambda_{Cl^-}} = \frac{t_{Ca^{2+}/CaCl_2}}{t_{Cl^-/CaCl_2}} \tag{4.194}$$

and substituting λ_{Cl^-} from Eq. (4.193),

$$\frac{\lambda_{Ca^{+2}}}{t_{Cl^-/KCl} \Lambda_{KCl}} = \frac{t_{Ca^{2+}/CaCl_2}}{t_{Cl^-/CaCl_2}} \tag{4.195}$$

Since $t_{Cl^-/KCl} = 1 - t_{K^+/KCl}$ and $t_{Cl^-/CaCl_2} = 1 - t_{Ca^{+2}/CaCl_2}$

$$\frac{\lambda_{Ca^{+2}}}{\left(1 - t_{K^+/KCl}\right)\Lambda_{KCl}} = \frac{t_{Ca^{2+}/CaCl_2}}{1 - t_{Ca^{+2}/CaCl_2}} \tag{4.196}$$

or

$$\lambda_{Ca^{+2}} = \frac{\left(1 - t_{K^+/KCl}\right) t_{Ca^{2+}/CaCl_2} \Lambda_{KCl}}{1 - t_{Ca^{+2}/CaCl_2}}$$

$$= \frac{(1 - 0.490)(0.438)\Lambda_{KCl}}{1 - 0.438} = 0.398\,\Lambda_{KCl}$$

(4.197)

Also, from Eq. (4.193),

$$\lambda_{Cl^-} = t_{Cl^-/KCl}\Lambda_{KCl} = \left(1 - t_{K^+/KCl}\right)\Lambda_{KCl}$$
$$= (1 - 0.490)\Lambda_{KCl} = 0.510\,\Lambda_{KCl}$$

(4.198)

and from Eq. (4.192)

$$\lambda_{K^+} = t_{K^+/KCl}\Lambda_{KCl} = 0.490\,\Lambda_{KCl}$$

(4.199)

With these values, now it is possible to calculate the transport number of Ca^{+2} in the mixture. Substituting Eqs. (4.197)-(4.199) into Eq. (4.188),

$$t_{Ca^{+2}\,mixture} =$$

$$= \frac{|2|(0.001M)(0.398\Lambda_{KCl})}{|2|(0.001M)(398\Lambda_{KCl}) + |1|(0.01M)(0.490\Lambda_{KCl}) + |1|(0.012M)(0.510\Lambda_{KCl})}$$

$$= 0.0673$$

(4.200)

where $c_{Cl^-} = 2c_{CaCl_2} + c_{KCl} = 0.012\,M$ was used.

Comment: Even if equivalent ionic conductivities are affected by concentration variation, this effect is lowered somehow by the transport numbers, which are calculated as ratios of λ's.

4.43 In Section 4.5.9 in the textbook the integration of the differential equation for the diffusion potential (Planck-Henderson equation) gave the following result (cf. Eq. 4.289 in the textbook)

$$-\Delta\psi = \frac{RT}{F}\sum \frac{t_i}{z_i} \ln\frac{c_i(l)}{c_i(0)}$$

(4.201)

Equation (4.201) was obtained considering (1) the activity coefficients were taken as unity, (2) the transport numbers were assumed constant, and (3) a linear variation of concentration with distance was assumed. If the second assumption is removed and the transport numbers are considered to vary with concentration as (cf. Section 4.5.2 in the textbook):

$$t_i = \frac{c_i u_i}{\sum_i c_i u_i} \qquad (4.202)$$

then the following equation for a 1:1 electrolyte would be obtained:

$$-\Delta \psi = \frac{RT}{F} \ln \frac{\sum_i c_i (l) u_i}{\sum_i c_i (0) u_i} \qquad (4.203)$$

(a) Prove Eq. (4.203) for a 1:1 electrolyte. (b) Calculate the junction potentials for the following situations: 0.1 M HCl/0.1 M KCl, and 0.1 M HCl / 0.01 M KNO$_3$, considering the following data:

Ion	u_{conv} (cm^2 s^{-1} V^{-1})	Ion	u_{conv} (cm^2 s^{-1} V^{-1})
H$^+$	3.625x10^{-3}	Cl$^-$	7.912x10^{-4}
K$^+$	7.619x10^{-4}	NO$_3^-$	7.404x10^{-4}

(Cf. Problem 4.16) (Kim)

Answer:

(a) The Planck-Henderson equation for diffusion potential is (cf. Eq. 4.284 in the textbook),

$$-\Delta \psi = \frac{RT}{F} \sum_i \frac{t_i}{z_i} d \ln a_i \qquad (4.204)$$

and the integral of this equation is (cf. Eq. 4.286 in the textbook),

$$-\Delta\psi = \frac{RT}{F}\sum_i \int_{x=0}^{x=l} \frac{t_i}{z_i} \frac{1}{f_i c_i} \frac{d(f_i c_i)}{dx}dx \qquad (4.205)$$

Considering an ideal solution, i.e., $f_i \approx 1$, Eq. (4.205) becomes

$$-\Delta\psi = \frac{RT}{F}\sum_i \int_{x=0}^{x=l} \frac{t_i}{z_i} \frac{1}{c_i} \frac{dc_i}{dx}dx \qquad (4.206)$$

Considering a linear variation of concentration with distance [cf. Fig. 4.86 in the textbook], i.e.,

$$c_i = c_i(0) + [c_i(l) - c_i(0)]x \qquad (4.207)$$

or

$$\frac{dc_i}{dx} = c_i(l) - c_i(0) \qquad (4.208)$$

Substituting Eq. (4.208) into Eq. (4.206),

$$-\Delta\psi = \frac{RT}{F}\sum_i \int_{x=0}^{x=l} \frac{t_i}{z_i} \frac{c_i(l) - c_i(0)}{c_i}dx \qquad (4.209)$$

Now, substituting c_i from (4.207) into the denominator of the equation for t_i, i.e., Eq. (4.202),

$$t_i = \frac{c_i u_i}{\sum_i c_i u_i} = \frac{c_i u_i}{\sum_i c_i(0)u_i + \sum_i [c_i(l) - c_i(0)]xu_i} \qquad (4.210)$$

Substituting Eq. (4.210) into Eq. (4.209), and considering $z_i = 1$,

$$-\Delta\psi = \frac{RT}{F}\sum_i \int_{x=0}^{x=l} \frac{c_i u_i}{\sum_i c_i(0)u_i + \sum_i [c_i(l) - c_i(0)]xu_i} \frac{c_i(l) - c_i(0)}{c_i}dx =$$

$$= \frac{RT}{F} \int\limits_{x=0}^{x=l} \frac{\sum\limits_{i}[c_i(l)-c_i(0)]u_i}{\sum\limits_{i}c_i(0)u_i + \sum\limits_{i}[c_i(l)-c_i(0)]xu_i}\, dx \qquad (4.211)$$

$$= \frac{RT}{F} \int\limits_{x=0}^{x=l} \frac{d\left\{\sum\limits_{i}c_i(0)u_i + \sum\limits_{i}[c_i(l)-c_i(0)]u_i x\right\}}{\sum\limits_{i}c_i(0)u_i + \sum\limits_{i}[c_i(l)-c_i(0)]xu_i}$$

$$= \frac{RT}{F} \ln\left\{\sum\limits_{i}c_i(0)u_i + \sum\limits_{i}[c_i(l)-c_i(0)]u_i x\right\}_{x=0}^{x=l}$$

Evaluating Eq. (4.211) between the given limits gives the equation we are looking for,

$$-\Delta\psi = \frac{RT}{F} \ln \frac{\sum\limits_{i}c_i(l)u_i}{\sum\limits_{i}c_i(0)u_i} \qquad (4.212)$$

(b) For 0.1 the M HCl/0.1 M KCl solution,

$$-\Delta\psi = \frac{RT}{F} \ln \frac{c_{K^+}u_{K^+} + c_{Cl^-}u_{Cl^-}}{c_{H^+}u_{H^+} + c_{Cl^-}u_{Cl^-}} = \frac{\left(8.314\,\mathrm{J\,mol^{-1}K^{-1}}\right)(298\,\mathrm{K})}{96500\,\mathrm{C\,mol^{-1}}}$$

$$\times \ln \frac{(0.1\,\mathrm{M})\left(7.619\times10^{-4}\,\mathrm{cm^2 s^{-1}V^{-1}}\right) + (0.1\,\mathrm{M})\left(7.912\times10^{-4}\,\mathrm{cm^2 s^{-1}V^{-1}}\right)}{(0.1\,\mathrm{M})\left(3.625\times10^{-3}\,\mathrm{cm^2 s^{-1}V^{-1}}\right) + (0.1\,\mathrm{M})\left(7.912\times10^{-4}\,\mathrm{cm^2 s^{-1}V^{-1}}\right)}$$

$$= -0.0266\,\mathrm{V}$$

$$(4.213)$$

In the same way for the 0.1 M HCl/0.01 M KNO$_3$ solution,

$$-\Delta\psi = \frac{RT}{F} \ln \frac{c_{K^+}u_{K^+} + c_{NO_3^-}u_{NO_3^-}}{c_{H^+}u_{H^+} + c_{Cl^-}u_{Cl^-}} = \frac{\left(8.314\,\mathrm{J\,mol^{-1}K^{-1}}\right)(298\,\mathrm{K})}{96500\,\mathrm{C\,mol^{-1}}} \times$$

$$\times \ln \frac{(0.01)\left(7.619\times10^{-4}\ \mathrm{cm}^2\mathrm{s}^{-1}\mathrm{V}^{-1}\right)+(0.01\mathrm{M})\left(7.404\times10^{-4}\right)}{(0.1)\left(3.625\times10^{-3}\ \mathrm{cm}^2\mathrm{s}^{-1}\mathrm{V}^{-1}\right)+(0.1\mathrm{M})\left(7.912\times10^{-4}\right)} = -0.0868\mathrm{V}$$

(4.214)

4.44 A two compartment electrochemical cell contains NaCl in one compartment and KCl in the other. The compartments are separated by a porous partition. Concentration of both electrolytes are equal. If Λ_{NaCl} and Λ_{KCl} are the equivalent conductivities of the two solutions, show that the liquid junction potential, E_L (or $-\Delta\psi$), is given by

$$E_L = \frac{RT}{F}\ln\frac{\Lambda_{NaCl}}{\Lambda_{KCl}}$$

(4.215)

(Contractor)

Answer:

The free energy change associated with the transfer process when one mole of charge is passed, is given by, (cf. Eq. 4.283 in the textbook)

$$dG = -Fd\psi = \sum_i \frac{t_i}{z_i}d\mu_i = RT\sum_i \frac{t_i}{z_i}d\ln a_i$$

(4.216)

This free energy change results in an electrochemical potential known as the liquid-junction potential, E_L. Consider that the liquid junction consists of a continuous series of lamina. If the difference in activity on the two sides of the lamina gives rise to a difference in chemical potential, $d\mu_i$ the liquid-junction potential between the two faces of the lamina can be represented by dE_L. Then,

$$dE_L = -\frac{RT}{F}\sum_i \frac{t_i}{z_i}d\ln a_i$$

(4.217)

The total liquid-junction potential can be obtained by adding the dE_L across each lamina, i.e.,

$$E_L = \int dE_L = -\frac{RT}{F} \int\limits_{a_I}^{a_{II}} \sum_i \frac{t_i}{z_i} d\ln a_i \qquad (4.218)$$

where a_I and a_{II} are the activities in the two compartments I and II. If the mixing fraction x is the fraction of solution II at a particular lamina, then the fraction of solution I will be $1-x$, and x will vary between 0 and 1 going from pure I to pure II. If c_i' is the concentration of the i-th species in solution I and c_i'' is the respective concentration in solution II, then at a lamina where the mixing fraction is x,

$$c_i = c_i'(1-x) + c_i''x = c_i' + \left(c_i'' - c_i'\right)x \qquad (4.219)$$

The transport number of the ion in that lamina, t_i, is,

$$t_i = \frac{c_i u_i}{\sum\limits_i c_i u_i} = \frac{u_i\left[c_i' + \left(c_i'' - c_i'\right)x\right]}{\sum\limits_i c_i' u_i + x\sum\limits_i u_i\left(c_i'' - c_i'\right)} \qquad (4.220)$$

where u_i is the mobility. Substituting concentrations for activities, the expression for liquid-junction potential becomes,

$$E_L = \int dE_L = -\frac{RT}{F} \int\limits_0^1 \sum_i \frac{1}{z_i} \frac{\left[c_i' + \left(c_i'' - c_i'\right)x\right]u_i}{\sum\limits_i c_i' u_i + x\sum\limits_i u_i\left(c_i'' - c_i'\right)} d\ln\left[c_i' + \left(c_i'' - c_i'\right)x\right]$$

$$= -\frac{RT}{F} \int\limits_0^1 \sum_i \frac{1}{z_i} \frac{\left[c_i' + \left(c_i'' - c_i'\right)x\right]u_i}{(1-x)\sum\limits_i c_i' u_i + x\sum\limits_i u_i c_i''} \frac{dx}{\left[c_i' + \left(c_i'' - c_i'\right)x\right]}$$

$$= -\frac{RT}{F} \int\limits_0^1 \sum_i \frac{1}{z_i} \frac{u_i}{(1-x)\sum\limits_i c_i' u_i + x\sum\limits_i u_i c_i''} dx \qquad (4.221)$$

or

$$E_L = -\frac{RT}{F} \frac{\sum_i \frac{u_i}{z_i}\left(c_i'' - c_i'\right)}{\sum_i u_i\left(c_i'' - c_i'\right)} \ln \frac{\sum_i u_i c_i'}{\sum_i u_i c_i''} \qquad (4.222)$$

For the system under consideration, I contains NaCl and II contains KCl. Considering also that the concentrations in the two compartments are equal then, Eq. (4.221) becomes,

$$E_L = -\frac{RT}{F} \frac{\left(u_{K^+} - u_{Cl^-}\right) - \left(u_{Na^+} - u_{Cl^-}\right)}{\left(u_{K^+} + u_{Cl^-}\right) - \left(u_{Na^+} + u_{Cl^-}\right)} \ln \frac{u_{Na^+} + u_{Cl^-}}{u_{K^+} + u_{Cl^-}} \qquad (4.223)$$

or

$$E_L = \frac{RT}{F} \ln \frac{\Lambda_{NaCl}}{\Lambda_{KCl}} \qquad (4.224)$$

Equation (4.222) is the needed expression.

MICRO-RESEARCH PROBLEMS

4.45. In an electrochemical experiment, a 10^{-3} M uni-valent ion solution is constantly oxidized at the electrode in the presence of a large amount of indifferent electrolyte. A current density of 1.0 mA cm^{-2} passes through the anode constantly. The diffusion coefficient of the ion is 10^{-9} m^2 s^{-1} at 25 ^{0}C. (a) Calculate the concentration of the ion at 0.05 seconds after the current is switched on, and at distances of 1 μm, 5.5 μm and 20 μm from the electrode. Now calculate the concentration of the ion at 0.50 seconds after the current is switched on and at distances of 25 μm, 40 μm and 100 μm from the electrode. Compare these results with those obtained at 0.05 seconds. (b) Based on the above results, as well as the results from Problem 4.30, draw, qualitatively, the three-dimensional distribution of the ions with respect to both, the time, t, and the distance x from the electrode. (c) Draw a similar three-dimensional distribution of the ions for an instantaneous pulse-induced diffusion problem. (Cf. Micro Research 4.1 in the textbook) (Xu)

Answer:

(a) The variation of the concentration of diffusing species with the distance x from the electrode/solution interface with the time that has elapsed since a constant consumption of ions is switched on is given by (cf. Eq. 4.72 in the textbook),

$$c = c^0 - \frac{\lambda}{\sqrt{D}} \left[2\sqrt{\frac{t}{\pi}} \, exp\left(-\frac{x^2}{4Dt} \right) - \frac{x}{\sqrt{D}} \, ercf \sqrt{\frac{x^2}{4Dt}} \right] \tag{4.225}$$

The flux, which is constant, is given by

$$\vec{j} = \frac{i}{zF} = \frac{0.001\,A\,cm^2}{96500\,C\,mol^{-1}} = 1.036 \times 10^{-8} \; mol\,cm^2\,s^{-1} \tag{4.226}$$

At a distance of $1\,\mu m$ from the electrode, the concentration of the ion after 0.05 s, is calculated from Eq. (4.223) with the following values for the corresponding variables: $t = 0.05$ s, $x = 1.0 \times 10^{-6}$ m, $j = 1.036 \times 10^{-4}$ mol m^{-2} s^{-1}, $c = 1$ mol m^{-3}, and $D = 1.0 \times 10^{-9}$ m^2 s^{-1}:

$$c\left[t = 0.05\,s, \; x = 10^{-6}\,m \right] = 1\,mol\,m^{-3} - \frac{1.036 \times 10^{-4} \; mol\,m^2\,s^{-1}}{\sqrt{10^{-9}\,m^2\,s^{-1}}}$$

$$\times \left\{ 2\sqrt{\frac{0.05\,s}{\pi}} exp\left[-\frac{10^{-12}\,m}{4\left(10^{-9}\,m^2\,s^{-1}\right)(0.05s)} \right] \right. \tag{4.227}$$

$$\left. -\frac{10^{-6}\,m}{\sqrt{10^{-9}\,m^2\,s^{-1}}} ercf \sqrt{\frac{10^{-12}\,m}{4\left(10^{-9}\,m^2\,s^{-1}\right)(0.05s)}} \right\}$$

According to Table 4.6 in the textbook,

$$erfc(0.071) = 1 - erf(0.071) = 1 - 0.0789 = 0.9211 \tag{4.228}$$

then,

$$c\left[t = 0.05\,s, x = 10^{-6}\,m\right] = 0.274\,mol\,m^{-3} \qquad (4.229)$$

In the same way for the other concentrations at $t = 0.05$ s and $t = 0.5$ s,

t = 0.05 s		t = 0.5 s	
x (m)	c (mol m⁻³)	x (m)	c (mol m⁻³)
1.0×10^{-6}	0.274	2.5×10^{-5}	0.114
5.5×10^{-6}	0.616	4.0×10^{-5}	0.671
2.0×10^{-5}	0.986	1.0×10^{-4}	0.982

Comments: (1) The concentration is closer to the initial value as the distance from the electrode surface increases. (2) At a certain distance, concentration rapidly drops with time elapsing.

(b) From the initial and boundary conditions as well as the results of Problem 4.30, we know the following facts concerning the concentration change for the constant flux problem in a 3D space at a time t and distance x:

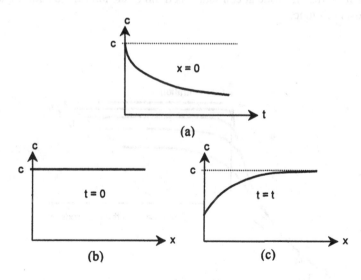

Figure 4.9. Variation of concentration with time and distance from the electrode for the continuous oxidation of an univalent ion.

(i) In the c-t plane at the electrode surface, $x = 0$, the concentration c varies as in Fig. 4.4, or Fig. 4.9(a).

(ii) When the time becomes equal to the transition time τ, the concentration becomes equal to zero at any distance from the electrode (cf. Problem 4.30c), i.e.,

$$c[t \geq \tau, x > 0] = 0 \qquad (4.230)$$

(iii) In the c-x plane, at any distance when $t = 0$, the concentration is equal to c^0, [cf. Fig. 4.9(b) and 4.9(c)], i.e.,

$$c[t = \tau, x = x] = c^0 \qquad (4.231)$$

At any time and at a distance far away from the electrode,

$$c[t = t, x \gg 0] = c^0 \qquad (4.232)$$

(iv) At any time after the oxidation starts, the slope of the c-x relationship at $c = 0$ is the same.

With this information the 3D map of concentration varying with time and distance from the electrode at constant flux drain of the ion can be built. Figure 4.10 shows this map.

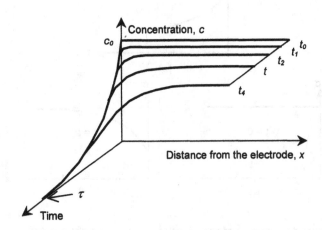

Figure 4.10. A c-t-x graph for a constant flux drain of an ion.

(c) Instantaneous pulse-induce diffusion. From Problem 4.30 and with the concentration variation with distance from the electrode described in Fig. 4.31 in the textbook, the following facts are known:

(i) At $t = 0$, the concentration everywhere but at the electrode surface is zero, i.e.,

$$c\left[t = 0, x = x\right] = 0 \qquad (4.233)$$

(ii) In the c-t plane at the surface of the electrode, the concentration varies as in Fig. 4.4 or Fig. 4.11(a).

(iii) At any time greater than zero, the relation c-x is shaped like a semi-bell, as in Fig. 4.31 in the textbook or Fig. 4.11(b)

(iv) At $t = \infty$ the concentration gradient disappears.

(v) At distance far from the surface of the electrode, the concentration of the ion is zero, i.e.,

$$c\left[t = t, x \gg 0\right] = 0 \qquad (4.234)$$

Figure 4.12 depicts the 3D map of concentration varying with time and distance with instantaneous pulse-induce diffusion.

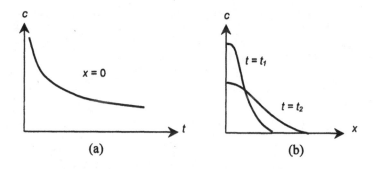

(a)   (b)

Figure 4.11. Variation of concentration with time and distance from the electrode for instantaneous pulse-induced diffusion.

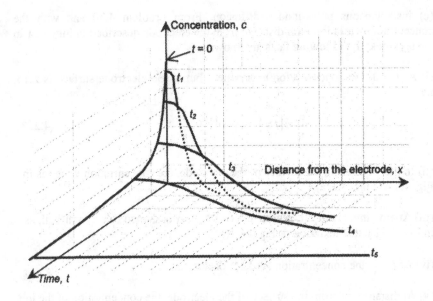

Figure 4.12. A *c-t-x* graph for instantaneous pulse-
induced diffusion.

CHAPTER 5

IONIC LIQUIDS

EXERCISES

Review of Sections 5.1 and 5.2 of the Textbook

Define *pure-liquid electrolyte*. Discuss similarities and differences between molten ice (i.e., water) and molten sodium chloride. What are *molten oxides*? Name similarities and differences between *pure-liquid electrolytes* and *aqueous solutions*. What do X-rays and nuclear diffraction tell about structural differences between fused salts and their corresponding crystals? How are the ions distributed in a fused salt as compared to an ionic crystal? Describe the diffraction-grating apparatus developed by Von Laue. Explain the meaning of *refraction* and *diffraction*. Write Bragg's equation describing each of the variables involved in the equation. Draw radial distribution functions as a function of distance, and explain the meaning of the maxima and minima in the plots. What does a pair-correlation function represent? How can pair-correlation functions be used to determine properties of liquids? Why is neutron diffraction preferred to X-rays in carrying out diffraction experiments with molten salts? What are the disadvantages of using neutrons in studying fused salts? What is the effect of *voids* in liquid salts on the pair-correlation functions? How do computer approaches help to determine properties of ions in liquids and solutions?

5.1 X-ray data on the internuclear distances in the solid and liquid forms of alkali-halides are given in the textbook. In all cases, the internuclear

distance *contracts* on melting. On the other hand, the volume of the solid lattice *increases* when it becomes liquid. Using the data of internuclear distances, find out the average contraction volume and compare it with the average increase in volume of the lattice upon melting. What kind of structure of molten salts does this suggest? (Cf. Exercises 5.23 and 5.24 in the textbook) (Bockris-GamboaAldeco)

Answer:

Table 5.9 in the textbook gives values of the internuclear distance of crystals and the corresponding molten salts. For example, for LiCl the internuclear distance in the ionic crystal is 266 pm, and that in the molten salt is 247 pm. Since the volume is proportional to the cubic distance, then, the change of volume upon melting is,

$$\% \Delta V = \frac{r_{liq}^3 - r_{crystal}^3}{r_{crystal}^3} x100 = \frac{247^3 - 266^3}{266^3} x100 = -19.9\% \qquad (5.1)$$

In the same way for the other salts given in Table 5.9 in the textbook,

Salt	Internuclear distances (pm)		% decrease in volume
	Crystal	Molten salt	
LiCl	266	247	19.9
LiBr	285	268	16.8
LiI	312	285	23.8
NaI	335	315	16.9
KCl	326	310	14.0
CsCl	357	353	3.3
CsBr	372	355	13.1
CsI	394	385	6.7
			Average: 14.3%

The average *contraction* volume for these salts considering the decrease of their internuclear distances is **14.3%**. The average increase in volume upon fusion for some salts can be obtained from the data in Table 5.10 in the textbook:

Salt	% increase of volume on fusion
NaCl	25
NaF	24
NaI	19
KCl	17
KBr	17
KI	16
RbCl	14
$CdCl_2$	20
$CdBr_2$	28
$NaNO_3$	11
	Average: 19.1%

The average *increase* in volume of the salts upon fusion is **19.1%**. These results indicate that the distances between atoms in the fused salts become smaller as compare to the corresponding crystals, and thus, the total volume of the sample should decrease by about the same amount. However, the total volume of the sample increases by 19.1%, indicating that some volume is not occupied by ions, and should correspond to free space in the sample.

Review of Sections 5.3 to 5.5 of the Textbook

Write the Born-Huggins-Meyer equation for pair-wise addition potentials for molten salts and explain each one of its terms. What is the contribution of Woodcock and Singer to the molten-salt theory? What are the factors involved in Saboungi's calculation? What are the improvements of the Saboungi's model over previous models? How is the much later work of Saboungi considered an improvement to the pioneer calculations of Woodcock and Singer? Explain the *hole model* for fused salts. How are the holes produced? What does a *distribution function* for the hole sizes represent? What is the *breathing motion* of a hole? Which are the variables needed to specify a hole and how are they different from those needed to specify an ion? Explain the meaning of *mass of a hole*. What is the contribution of Fürth to fused-salt theory? Write the Fürth equation for the work done in expanding a hole. Starting from the equation of the probability of the location, momentum and breathing momentum of a hole as a function of the Boltzmann-probability factor, derive the average radius of a hole in fused salts. What are the conclusions arising from Fürth's theory of holes in liquids? What are *supercooled liquids*? What is *glass-transition temperature*? What is the contribution of Angell and of Cohen and Turnbull to molten-salt theory? Describe the property of *free volume* and write a corresponding equation for this variable. How does the equation of probability of finding a hole in the

Cohen-Turnbull model differ from that in the Fürth's model? What is the available proof for the validity of the Cohen-Turnbull theory of glass-forming molten salts?

5.2 Calculate the work of hole formation in molten sodium chloride using the Fürth approach. Take the surface tension of NaCl salt at 900 ^{0}C as 107.1 dyne cm^{-1}, and the mean-hole radius of NaCl as 1.7x10^{-8} cm. (Cf. Exercise 5.10 in the textbook) (Contractor)

Answer:

The work of hole formation can be represented as (cf. Eq. 5.33 in the textbook),

$$W = 4\pi r^2 \gamma \tag{5.2}$$

where r and γ are the mean-hole radius and the surface tension, respectively. Therefore,

$$W = 4\pi \left(1.7 \times 10^{-8}\ cm\right)^2 \left(107.1\, dyne\ cm^{-1}\right) \times \frac{1\,erg}{1\,dyne\ cm^{-1}} = 3.89 \times 10^{-13}\ erg$$

$$= 3.89 \times 10^{-20}\ J \tag{5.3}$$

or **23.4 kJ mol^{-1}**.

5.3 What is the hole radius of maximum probability found in molten sodium chloride at 900 ^{0}C? The surface tension of molten sodium chloride at this temperature is 107.1 dyne cm^{-1}. Compare your calculated value with the mean-hole radius of the salt. (Contractor)

Answer:

The probability of the existence of a hole of radius r is given by (cf. Eq. 5.39 in the textbook),

$$P_r\, dr = \frac{16}{15\pi^{1/2}} a^{7/2} r^6 e^{-ar^2}\, dr \tag{5.4}$$

where

$$a = \frac{4\pi\gamma}{kT} \qquad (5.5)$$

The distribution function given in Eq. (5.4) can be used to plot the probability P_r against the radius of the hole. The maximum value of P_r in this plot corresponds to the most probable size of the hole found in the sample. To make the graph, r can take values from, say, 1.0×10^{-8} cm to 3.0×10^{-8} cm at intervals of 0.2×10^{-8} cm. With this in mind, Eq (5.4) can be evaluated. The value of a from Eq. (5.5) is,

$$a = \frac{4\pi\left(107.1\,\text{dyne cm}^{-1}\right)}{\left(1.38 \times 10^{-16}\,\text{erg K}^{-1}\right)\left(1173\,\text{K}\right)} \times \frac{1\,\text{erg}}{1\,\text{dyne cm}} = 8.314 \times 10^{15}\,\text{cm}^{-2} \quad (5.6)$$

The value of P_r at 1.0×10^{-8} cm is

$$P_r = \frac{16}{15\pi^{1/2}}\left(8.314 \times 10^{15}\,\text{cm}^{-2}\right)^{7/2}\left(1.0 \times 10^{-8}\,\text{cm}\right)^6$$
$$\times e^{-\left(8.314 \times 10^{15}\,\text{cm}^{-2}\right)\left(1.0 \times 10^{-8}\,\text{cm}\right)^2} = 1.373 \times 10^7\,\text{cm}^{-1} \qquad (5.7)$$

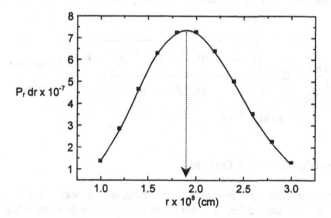

Figure 5.1. Probability of the existence of a hole of radius r in molten NaCl.

In the same way for the other values of r,

$r \times 10^8$ (cm)	$P_r \times 10^{-7}$	$r \times 10^8$ (cm)	$P_r \times 10^{-7}$
1.0	1.373	2.2	6.392
1.2	2.843	2.4	5.014
1.4	4.654	2.6	3.529
1.6	6.296	2.62	3.388
1.8	7.250	2.8	2.243
1.9	7.375	3.0	1.293
2.0	7.255	3.8	0.058

A corresponding plot of the probability against the radius of the ion is given in Fig. 5.1. The maximum in the plot of P_r vs. r occurs at 1.9×10^{-8} cm, which represents the most probable radius of the holes in molten NaCl. This value is comparable to the mean-hole radius, 1.7×10^{-8} cm given in the textbook in Table 5.15.

5.4 Calculate the mean-hole size of fused CsBr for which the surface tension is 60.7 dynes cm^{-1} at 900 ^{0}C. (Cf. Exercise 5.3 in the textbook) (Bockris-GamboaAldeco)

Answer:

The mean–hole radius in molten salts is given by (cf. Eq. 5.44 in the textbook),

$$\langle r \rangle = 0.51 \sqrt{\frac{kT}{\gamma}} = 0.51 \sqrt{\frac{\left(1.381 \times 10^{-16}\ \text{ergK}^{-1}\right)\left(1173\,\text{K}\right)}{60.7\,\text{dyne cm}^{-1}} \times \frac{1\,\text{dyne cm}}{1\,\text{erg}}} \qquad (5.8)$$

$$= 2.63 \times 10^{-8}\ \text{cm}$$

Review of Section 5.6 of the Textbook

How are the transport phenomena of diffusion as well as the ion-ion interactions in fused salts different from those in aqueous solutions? What is the meaning of *self-diffusion*? How is it measured? What is the order of magnitude of self-diffusion of Na$^+$ in molten NaCl compared to that in crystal NaCl near the melting point? Write empirical expressions for self-diffusion as a function of temperature. Write an equation for the activation energy for self-diffusion. Write similar equations for the viscosity of molten salts. What is the main conclusion due to the general applicability of the Nanis-Bockris equation? Write the Stokes-Einstein equation for diffusion. Does this equation apply successfully to ions in molten salts? Why has the electrical conductivity of fused salts been widely studied? What general trends referring to the size and valence of the ions can be withdrawn from the equivalent conductivities of ions in molten salts? Do experimental data for ionic liquids fit the Nernst-Einstein equation properly? Under what conditions would the conductivity exceed the observed value? Discuss the diffusion mechanism proposed by Borucka. Derive an equation explaining the deviation from Nernst-Einstein equation in fused salts. Define *transport number*. What are the variables used to measure transport numbers in fused salts? Describe some experiments by which this is done. Explain why the direct determination of transport numbers in molten slats is more difficult in comparison with the same determination in aqueous electrolytes. Would it not be better to abandon them and use the approximate applicability of the Nernst-Einstein equation relying on self-diffusion determinations?

5.5 The melting temperature of $CaCl_2$ is 919 K and the pre-exponential factor for self-diffusion of ^{36}Cl is 1.8×10^{-3} cm^2 s^{-1} and of ^{45}Ca is 0.38×10^{-3} cm^2 s^{-1}. Given these data, determine the self-diffusion coefficient of both ions in the salt at 1200 K. Establish any needed assumption. (GamboaAldeco)

Data:

$T_{m.p.} = 919$ K $\qquad\qquad\qquad D_{0,Cl} = 1.8 \times 10^{-3}$ cm^2 s^{-1}
$T = 1200$ K $\qquad\qquad\qquad D_{0,Ca} = 0.38 \times 10^{-3}$ cm^2 s^{-1}

Answer:

The diffusion coefficient can be obtained from (cf. Eq. 5.55 in the textbook)

$$D = D_0 \, exp\left(-\frac{E_D^{\neq}}{RT} \right) \qquad\qquad (5.9)$$

Considering that the value of the activation energy is similar for both ions, and that it does not change in this range of temperature, the Nanis-Bockris equation at 919 K reads (cf. Eq. 5.53 in the textbook):

$$E_D^{\neq} = 3.74\, RT_{m.p.} = 3.74\left(8.314\,\text{J mol}^{-1}\text{K}^{-1}\right)(919\,\text{K})$$

$$= 28575\,\text{J mol}^{-1}$$

(5.10)

For $^{45}Ca^{+2}$,

$$D_{Ca^{+2}} = 0.38 \times 10^{-3}\,\text{cm}^2\text{s}^{-1}\ exp\left(-\frac{28575\,\text{J mol}^{-1}}{\left(8.314\,\text{J mol}^{-1}\text{K}^{-1}\right)(1200\,\text{K})}\right)$$

(5.11)

$$= 2.17 \times 10^{-5}\,\text{cm}^2\text{s}^{-1}$$

and for $^{36}Cl^-$,

$$D_{Cl^-} = \left(1.8 \times 10^{-3}\,\text{cm}^2\text{s}^{-1}\right) exp\left(-\frac{28575\,\text{J mol}^{-1}}{\left(8.314\,\text{J mol}^{-1}\text{K}^{-1}\right)(1200\,\text{K})}\right)$$

(5.12)

$$= 10.3 \times 10^{-5}\,\text{cm}^2\text{s}^{-1}$$

5.6 The diffusion coefficient of the tracer Cl⁻ ion in molten NaCl is 4.2x10⁻⁵ cm² s⁻¹ at 1020 °C, and 6.7x10⁻⁵ cm² s⁻¹ at 840 °C. (a) Calculate the values of the activation energy and the pre-exponential factor for the diffusion process. (b) Compare the $E^{\neq}{}_D$ value with that obtained from the empirical equation proposed by Nanis-Bockris, i.e., $E^{\neq}{}_D = 3.74\ RT_{m.p.}$. (Cf. Exercise 5.33 in the textbook) (Contractor)

Answer:

(a) The self-diffusion coefficient is given by (cf. Eq. 5.55 in the textbook),

$$D = D_0\ exp\left(\frac{-E_D^{\neq}}{RT}\right)$$

(5.13)

The activation energy, $E^{\neq}{}_D$, and the pre-exponential factor, D_0, can be obtained from the solution of two simultaneous equations given by Eq. (5.13) at two different temperatures, i.e., $ln\, D_1 = ln\, D_0 - E^{\neq}_D/RT_1$ and ., $ln\, D_2 = ln\, D_0 - E^{\neq}_D/RT_2$. Subtracting these two equations,

$$ln\, D_1 - ln\, D_2 = -\frac{E^{\neq}_D}{R}\left(\frac{1}{T_1} - \frac{1}{T_2}\right) \tag{5.14}$$

and substituting the appropriate values,

$$ln\, 6.7 x 10^{-5} - ln\, 4.2 x 10^{-5} = -\frac{E^{\neq}_D}{8.314\,J\,K^{-1}mol^{-1}}\left(\frac{1}{1113K} - \frac{1}{1293K}\right) \tag{5.15}$$

Solving for $E^{\neq}_D$ gives, $\mathbf{E^{\neq}_D = 31\ kJ\ mol^{-1}}$. Substituting this value of $E^{\neq}_D$ in Eq. (5.13), and solving for $ln\, D_0$:

$$ln D_0 = ln D_1 + \frac{E^{\neq}_D}{RT_1} = ln\, 6.7 x 10^{-5} + \frac{31 x 10^3\, J\, mol^{-1}}{\left(8.314\ J\,K^{-1}\,mol^{-1}\right)(1113K)} = -6.26 \tag{5.16}$$

Equation (5.16) gives a value for the pre-exponential factor for this salt of $D_0 = 1.9 x 10^{-3}\ cm^2\ s^{-1}$.

(b) The melting point of NaCl can be obtained from Table 5.6 in the textbook. Therefore,

$$E^{\neq}_D = 3.74\, RT_{m.p.} = 3.74\left(8.314\,J\,K^{-1}\,mol^{-1}\right)(1074K) = \mathbf{33.4\,kJ\,mol^{-1}} \tag{5.17}$$

This value is in good agreement with that found from experimental data of diffusion coefficients.

5.7 It is known that the equivalent conductivity of molten salts depends on the size of the cation forming the salt. Using the data in the book, plot the equivalent conductivities of molten salts of monovalent cations against the radii of the corresponding cations. Which type of dependence did you find? (Cf. Exercise 5.32 in the textbook) (Contractor)

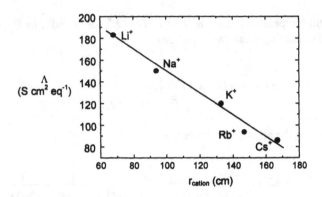

Figure 5.2. Equivalent conductivity of molten salts of monovalent cations as a function of the cation radii.

Answer:

The next table (cf. Table 5.24 in the textbook) shows the equivalent conductivities of molten chlorides and the radii of the corresponding cations, and Figure 5.2 shows the corresponding plot:

	LiCl	NaCl	KCl	RbCl	CsCl
Radius of the cation (pm)	68	94	133	147	167
Equivalent conductivity (S cm^2 eq^{-1})	183	150	120	94	86

The equivalent conductivities of monovalent cations show a linear dependence with respect to the corresponding ionic radius.

Comment: The larger ions such as Cs$^+$ and Rb$^+$ need a larger hole to jump into to be able to move. Consequently, their conductivities are smaller than those of the smaller cations.

5.8 The equivalent conductivity of molten NaCl is 153 Ω^{-1} cm^2 eq^{-1} at 1153 K, and the self-diffusion coefficients of Na$^+$ and Cl$^-$ ions in molten NaCl are 9.6x10^{-5} and 6.7x10^{-5} cm^2 s^{-1}, respectively. With these data evaluate Faraday's constant. (Cf. Exercise 5.34 in the textbook) (Contractor)

Answer:

The Faraday's constant can be calculated using the Nernst-Einstein relation (cf. Eq. 5.61 in the textbook),

$$\Lambda = \frac{F^2}{RT}\left(D_{Na^+} + D_{Cl^-}\right) \qquad (5.18)$$

or,

$$F = \sqrt{\frac{\Lambda RT}{D_{Na^+} + D_{Cl^-}}} = \sqrt{\frac{\left(153\,\Omega^{-1}\,cm^2\,eq^{-1}\right)\left(8.314\,J\,eq^{-1}\,K^{-1}\right)(1153)}{(9.6 + 6.7) x 10^{-5}\,cm^2\,s^{-1}}}$$

$$\times \sqrt{\frac{1C^2\,\Omega s^{-1}}{1J}} = 94857 \ C\ eq^{-1} \qquad (5.19)$$

Review of Section 5.7 of the Textbook

Write a molecular-kinetic expression for *viscosity* of a fluid and explain the model of *viscous flow* in fluids. How is the viscosity represented under this model? How is this molecular-kinetic expression modified for ionic liquids? Explain the significance of the *mean-time between collisions* in an ionic liquid and write and equation for it. Write an expression for the viscosity of ionic liquids as a function of the mean radius of the holes. Write the Stokes-Einstein relation for the diffusion coefficient of holes. Starting from this relation, derive an expression for the diffusion coefficients of ions in a fused salt and as function of the volume expansion at the melting point. Derive Eq. (5.101) in the textbook. Derive an expression for the heat of activation of *hole formation, A*, as a function of the melting temperature of the salt. What are the assumptions considered in this derivation? Mention some of the advantages and difficulties the hole model has. Why is it said that the variation of the diffusion coefficient with temperature at *constant volume* is an artificial state? How are the heats of activation for *hole formation* and that for *jumping-into-into-a-hole* determined? How does Swallin's model differ from the *jumping-into-holes* model? Why is this model rejected as a possible model to explain conductivity in molten salts? Explain Rice and Allnutt's model for molten salts and mention some of its drawbacks.

5.9 For a hole in molten KCl at 1100 K, calculate (a) its average radius, (b) its mass, and (c) its lifetime. The melting point of this salt is 1045 K, and in its liquid state, its surface tension is 89.5 dyn cm^{-1}, and its molar volume is 50.20 cm^3 mol^{-1}. (Cf. Exercise 5.19 in the textbook) (Bockris-GamboaAldeco)

Data:

$T = 1100$ K $V_{m,KCl} = 50.20$ cm^3 mol^{-1} $\gamma_{KCl} = 89.5$ dyn cm^{-1}

$T_{m.p.} = 1045$ K $M_{KCl} = 74.6$ g mol^{-1}

Answer:

(a) The average-hole radius is given by (cf. Eq. 5.44 in the textbook):

$$\langle r_h \rangle = 0.51 \left(\frac{kT}{\gamma} \right)^{1/2} = 0.51 \sqrt{\frac{\left(1.381 \times 10^{-23} \text{ J K}^{-1} \right)\left(1100 \text{K} \right)}{89.5 \times 10^{-3} \text{ N m}^{-1}}} \times \frac{1 \text{N m}}{1 \text{J}} \quad (5.20)$$

$$= 2.10 \times 10^{-10} \text{ m}$$

(b) The apparent mass of a hole is given by (cf. Eq. 5.28 in the textbook),

$$m_h = \frac{2}{3} \pi \langle r_h \rangle^3 \rho \quad\quad\quad (5.21)$$

where ρ is the density of the liquid. Therefore,

$$\rho = \frac{1}{V_{m,KCl}} = \frac{1 \text{mol}}{50.20 \text{cm}^3} \times \frac{0.0746 \text{ kg}}{1 \text{mol}} \times \frac{10^6 \text{ cm}^3}{1 \text{m}^3} = 1486 \text{ kg m}^{-3} \quad (5.22)$$

Substituting Eqs. (5.20) and (5.22) into Eq. (5.21),

$$m_h = \frac{2}{3} \pi \left(2.10 \times 10^{-10} \text{ m} \right)^3 \left(1486 \text{ kg m}^{-3} \right) = 2.88 \times 10^{-26} \text{ kg} \quad (5.23)$$

or 17.4 g mol^{-1}.

(c) The lifetime of a hole is calculated by (cf. Eq. 5.93 in the textbook)

$$\tau = \frac{\langle r_h \rangle}{3} \sqrt{\frac{2\pi m_h}{kT}} e^{A/RT} \tag{5.24}$$

where (cf. Eq. 5.115 in the textbook)

$$A = 3.3 \, RT_{m.p.} = 3.3(8.314 \text{ J K}^{-1} \text{ mol}^{-1})(1045 \text{ K}) = 28670 \text{ J mol}^{-1} \tag{5.25}$$

Substituting Eqs. (5.20), (5.23) and (5.25) into Eq. (5.24),

$$\tau = \frac{\left(2.10 \times 10^{-10} \text{ m}\right)}{3} \sqrt{\frac{2\pi\left(2.88 \times 10^{-26} \text{ kg}\right)}{\left(1.38 \times 10^{-23} \text{ J K}^{-1}\right)(1100 \text{ K})}}$$

$$\times \exp\left[\frac{28670 \text{ J mol}^{-1}}{\left(8.314 \text{ J K}^{-1} \text{mol}^{-1}\right)(1100 \text{K})}\right] \times \frac{1 \text{ J}^{1/2}}{1 \text{ kg}^{1/2} \text{ m s}^{-1}} \tag{5.26}$$

$$= 5.56 \times 10^{-12} \text{ s}$$

5.10 Deduce an expression to calculate the number of holes per unit volume of liquid in a molten salt in terms of its physical properties (e.g., temperature, density). (GamboaAldeco)

Answer:

The number of holes per unit volume of the liquid, n_h, can be obtained from the expression for viscosity of the hole-theory (cf. Eq. 5.94 in the textbook),

$$\eta = \frac{2}{3} n_h \langle r_h \rangle (2\pi m k T)^{1/2} e^{A/RT} \tag{5.27}$$

The average radius of the hole is (cf. Eq. 5.44 in the textbook),

$$\langle r_h \rangle = \frac{8}{5\pi} \left(\frac{kT}{\gamma}\right)^{1/2} \tag{5.28}$$

The parameter A is given by [cf. Section5.7.6 in the textbook],

$$A = 3.3 \, RT_{m.p.} \tag{5.29}$$

and the mass of a hole of average radius is (cf. Eq. 5.28 in the textbook).

$$m_h = \frac{2}{3}\pi \langle r_h \rangle^3 \rho \tag{5.30}$$

Substituting these expressions into equation (5.27) and solving it for n_h,

$$
n_h = \frac{3}{2}\frac{\eta e^{-A/RT}}{\langle r_h \rangle (2\pi m_h kT)^{1/2}} = \frac{3}{2}\frac{\eta e^{-A/RT}}{(2\pi kT)^{1/2}}\left(\frac{3}{2}\frac{1}{\pi\rho}\right)^{1/2}\frac{1}{\langle r_h \rangle^{5/2}} =
$$

$$
= \left(\frac{27}{16}\right)^{1/2}\frac{\eta e^{-A/RT}}{\pi (kT\rho)^{1/2}}\left(\frac{5\pi}{8}\right)^{5/2}\left[\left(\frac{\gamma}{kT}\right)^{5/2}\right]^{1/2} \tag{5.31}
$$

Arranging terms

$$\mathbf{n_h} = 2.234\sqrt{\gamma^{5/2}\eta^2 / (kT)^{7/2}}\,\rho\,e^{-3.3T_{m.p.}/T} \tag{5.32}$$

This equation gives the number of holes per unit volume, n_h, as a function of the surface tension, viscosity, density, temperature and melting point of the molten salt.

5.11 Using the equation developed in Exercise 5.10, calculate for NaCl fused salt at 1173 K, (a) the number of holes per cubic centimeter, (b) the number of holes per mole of fused salt. The melting point of NaCl is 1074.1 K. Its viscosity at its melting point is 1.67 cpoise, and at 1173 K, is 1.05 cpoise. The density of the salt at 1173 K is about 1.52 g cm^{-3}, and the surface tension at this temperature is 107.1 dyn cm^{-1}. (GamboaAldeco)

Data:

$T_{m.p.} = 1074$ K $\eta_{1074K} = 1.67$ cp $\rho_{1173} = 1.52$ g cm^{-3}

$T = 1173$ K $\eta_{1173K} = 1.05$ cp $\gamma_{1173K} = 107.1$ dyn cm^{-1}

Answer:

(a) The expression to calculate the number of holes per unit volume of liquid is (cf. Eq. 5.32 in Exercise 5.10),

$$n_h = 2.234 \sqrt{\frac{\gamma^{5/2} \eta^2}{(kT)^{7/2} \rho}} \, e^{-3.3 T_{m.p.}/T} \tag{5.33}$$

Substituting the appropriate values for fused NaCl at 1173 K,

$$n_h = 2.234 \, e^{-3.3(1074\,\text{K})/(1173\,\text{K})}$$

$$\times \sqrt{\frac{\left(107.1\,\text{dyn cm}^{-1}\right)^{5/2} (0.0105\,\text{poise})^2}{\left(1.381 \times 10^{-23}\,\text{J K}^{-1}\right)^{7/2} (1173\,\text{K})^{7/2} \left(1.52\,\text{g cm}^{-3}\right)}}$$

$$\times \sqrt{\left(\frac{1\,\text{g cm}^{-1}\text{s}^{-1}}{1\,\text{poise}}\right)^2 \left(\frac{1\,\text{J}}{10^7\,\text{dyn cm}}\right)^{7/2} \frac{1\,\text{dyn}}{1\,\text{g cm s}^{-2}}} \tag{5.34}$$

$$= 7.723 \times 10^{21} \, \frac{\text{holes}}{\text{cm}^3}$$

or,

$$n_h = 7.723 \times 10^{21} \, \frac{\text{holes}}{\text{cm}^3 \text{ of liquid}} \times \frac{1\,\text{mol}}{6.022 \times 10^{23}\,\text{holes}} \tag{5.35}$$

$$= 0.0128 \, \frac{\text{mol holes}}{\text{cm}^3}$$

(b) The number of holes per mole of fused salt is,

$$n_h \frac{M_{NaCl}}{\rho_{1173\,\text{K}}} = 0.0128 \frac{\text{mol holes}}{\text{cm}^3 \text{ of liquid}} \frac{58.5\,\text{g mol}^{-1}}{1.52\,\text{g cm}^{-3}} = 0.493 \frac{\text{mol holes}}{\text{mol salt}} \tag{5.36}$$

5.12 What fraction of the total volume is due to holes in the salt described in Exercise 5.10? Does the hole model conforms to the observed experimental increase of volume upon fusion? (GamboaAldeco)

Answer:

The volume of a hole of average radius is given by:

$$v_h = \frac{4}{3}\pi \langle r_h \rangle^3 = \frac{4}{3}\pi \left(\frac{8}{5\pi}\sqrt{\frac{kT}{\gamma}} \right)^3$$

$$= 0.553 \left[\frac{\left(1.38 \times 10^{-23}\ J\,K^{-1}\right)(1173K)}{107.1\,dyn\,cm^{-1}} \right]^{3/2} \left(\frac{10^7\ dyn\,cm}{1\,J} \right)^{3/2} \quad (5.37)$$

$$= 3.251 \times 10^{-23}\ cm^3$$

and the volume of holes in 1 cm^3 of liquid is

$$v_h n_h = \left(3.251 \times 10^{-23}\ \frac{cm^3}{hole} \right) \left(7.723 \times 10^{21}\ \frac{holes}{cm^3\ of\ liquid} \right) \quad (5.38)$$

$$= 0.251 \frac{cm^3\,hole}{cm^3}$$

The value obtained in Eq. (5.38) indicates that 25.1% of the total volume of the liquid is occupied by holes. Table 5.10 in the textbook shows an increase of volume of 25% upon fusion for NaCl. Therefore, the hole model used in this calculation conforms to the experimental behavior of the fused salt.

Review of Sections 5.8 and 5.9 of the Textbook

Draw a sketch of conductivity vs. composition for an *ideal* and *real* mixture of molten salts. In a general way, how are deviations from ideality explained in a mixture of molten salts? How is *a complexed ion* defined in molten salts? Define *transition time* and write an equation for the electrode potential as a function of the transition time. Explain a method based on transition times to determine the concentration of complexes in mixtures of fused salts. What is the parameter used to distinguish the existence of complexes in molten salts? What are some spectroscopic and electrochemistry methods used to determine complexes in molten salts? Describe the electrochemical process for the obtention of aluminum. What do the two peaks found in the Raman spectrum of cryolite represent? Explain how the coordination number is related to the value of the wavenumber of the given peak in a Raman spectrum. How is it possible that anions such as $AlOF^-$ react at electrodes negatively charged, i.e., cathodes? Which are the structures formed when $AlCl_3$ is added to $SnCl_2$? Describe briefly how Nuclear Magnetic Resonance works. What information related to molten salts is obtained from NMR? How is NMR used to study complexes of $AlCl_3$?

5.13 What is the force constant of vibration, k, of the entities shown in the Raman spectrum of molten Cryolite? The wavenumbers at which the peaks appear are 554 cm^{-1} and 575 cm^{-1}, respectively (cf. Fig. 5.57 of the textbook). Comment on your results. (Cf. Exercise 5.22 in the textbook) (GamboaAldeco)

Answer:

The frequency of a given vibration is given by (Eq. 2.24 in the textbook):

$$v = \frac{1}{2\pi}\sqrt{\frac{k}{\bar{\mu}}} \tag{5.39}$$

where $\bar{\mu}$ is the reduced mass given by (cf. Section 2.11.2 in the textbook)

$$\frac{1}{\bar{\mu}} = \frac{1}{m_1} + \frac{1}{m_2} = \frac{1}{M_{Al}} + \frac{1}{M_F} = \frac{1}{27} + \frac{1}{19} = \frac{1}{11.15\,g\,mol^{-1}} \tag{5.40}$$

and the frequency v is given by,

$$v = c_0\,\bar{v} \tag{5.41}$$

Therefore, from Eqs. (5.39)-(5.41),

$$k_{554\,cm^{-1}} = (2\pi v)^2\,\bar{\mu} = \left[2\pi\left(3 \times 10^{10}\,cm\,s^{-1}\right)\left(554\,cm^{-1}\right)\right]^2$$
$$\times \left(11.15\,g\,mol^{-1}\right)\frac{1\,mol}{6.022 \times 10^{23}} = 2.02 \times 10^5\,g\,s^{-2} \tag{5.42}$$

In the same way at 575 cm^{-1},

$$k_{575\,cm^{-1}} = 2.17 \times 10^5\,g\,s^{-2} \tag{5.43}$$

Comment: A high wavenumber indicates a large force constant. This means that the bond strength between the two considered atoms, e.g., atoms A and B, is stronger the higher the wavenumber is. If more atoms surround the central atom A (higher coordination number), the bond strength between A and B decreases, k becomes smaller, and thus the wavenumber decreases too.

Review of Section 5.10 of the Textbook

What is the effect of alkali metals dissolved in molten salts? Is the mobility of molten salts very different from that of the corresponding ions dissolved in aqueous solutions? What is the travelling lifetime of an electron in moving from one ion to another? Describe the three steps followed to calculate electronic conductance in molten salts according to Bockris and Emi.

5.14 Suppose the electrical conductivity of CaO is determined primarily by the diffusion of the Ca^{+2} ions. Estimate the mobility and the conductivity of this cation at 1800 °C. The diffusion coefficient of Ca^{+2} ion in CaO at this temperature is 10^{-14} m^2 s^{-1}. CaO has a NaCl structure with a parameter $a = 4.522$ Å. (Cf. Exercise 5.31 in the textbook) (Contractor)

Answer:

In ionic materials like CaO, the mobility of the charge carriers (or ions) is given by (cf. Eqs. 4.152 and 4.172 in the textbook),

$$u_e = u_{conv} = \frac{z_i e_0 D}{kT} = \frac{(2)\left(1.602 \times 10^{-19} \text{ C}\right)\left(10^{-14} \text{ m}^2 \text{s}^{-1}\right)}{\left(1.381 \times 10^{-23} \text{ JK}^{-1}\right)(2073 \text{K})} \frac{1\text{J}}{1\text{CV}} \quad (5.44)$$

$$= 1.12 \times 10^{-13} \text{ m}^2 \text{V}^{-1} \text{s}^{-1}$$

The conductivity is given by (cf. Eq. 5.134 in the textbook),

$$\sigma = FN_e u_e \quad (5.45)$$

The number of moles of electrons per cubic centimeter, N_e, is

$$N_e = \frac{number\ of\ Ca^{+2}\ ions\ per\ unit\ cell}{volume\ of\ unit\ cell} \times \frac{z}{N_A} = \frac{1}{\left(4.522 \times 10^{-10}\ m\right)^3}$$

$$\times \frac{2}{6.022 \times 10^{23}\ mol^{-1}} = 35910\left(mol\ of\ electrons\right)m^{-3} \quad (5.46)$$

Thus, the conductivity is calculated as,

$$\sigma = FN_e u_e = \left(96500\ C\ mol^{-1}\right)\left(35910\ mol\ m^{-3}\right)$$

$$\times \left(1.12 \times 10^{-13}\ m^2\ V^{-1} s^{-1}\right) \times \frac{1\ V}{1 C\ s^{-1}\ \Omega} = 3.88 \times 10^{-4}\ S\ m^{-1} \quad (5.47)$$

Comment: The mobility is many orders of magnitude lower than the mobility of electrons. Hence, the conductivity is very small compared to the electronic conduction.

Review of Sections 5.11 and 5.12 of the Textbook

Why is a liquid medium a better place for reactions to occur than a gaseous medium at the same temperature? Describe Lin's work and its importance in industry. Which is the *potential window* in which aqueous solutions can be examined before oxygen and hydrogen are evolved at low pH's? How is the potential window in, say, molten NaCl? Mention advantages of having systems with large potential windows. Is it possible to have the advantages of molten salts at room temperatures? Describe the acid-base behavior of low-melting-point liquid electrolytes such as those containing $AlCl_3$. Mention some onium salts. What are the structural characteristics of these salts? Explain why the melting point of molten salts decreases as the complexity of the molecule increases. Mention some disadvantages of the use of complex-molten salts. Why is it said that *protons act as contaminants in low-temperature-molten salts*?

5.15 Ambient-temperature molten salts are made up from certain alkyl-ammonium salts or alternatively, by a mixture of $AlCl_3$ with organic compounds. Mention two advantages of these salts over traditional molten salts. Suggest three solvents which would allow the electrochemical oxidation of complex organics such as polymerized isoprene (rubber) at less than 100 °C. (Cf. Exercise 5.28 in the textbook) (Bockris)

Answer:

Some of the advantages of ambient-temperature molten salts are : (1) They melt several hundred degrees below room temperature resulting in a material easier to handle. (2) They allow a much larger electrochemical window. Examples of these materials are: tetraheptylammonium chloride (m.p. 264 K), ethyl-diphenyl-sulfonium tetrafluoroborate (m.p. 308 K),and trimethyl-phenyl-phosphonium iodide (m.p. 308 K).

Review of Section 5.13 of the Textbook

What is an *oxide system*? Discuss the structural similarities between water and fused silica that make the conductivity of fused silica compounds lower than molten salts. How is the transport process in molten silica? Which is the rate determining process for transport in molten silica? What similarities and differences exist between water and fused silica? Between fused salts and fused silica? Explain the temperature dependence of E_η in liquid SiO_2. Which are the two ways water interacts with ions? How is the mechanism of dissolution of non-metallic oxides. Draw a schematic of this process. What is a *glass*? What is the effect of addition of metallic oxides to silica oxide? What is understood by *liquid silicate*, *frozen liquid*, and *devitrification* when referring to silicates? What happens when devitrification occurs? How is the *concentration* of metal oxides in nonmetallic oxides expressed? Draw a schematic of changes of activation energy for viscous flow as a function of the O/R ratio. What are the species present at a ratio O/Si > 4? Why does the theory involving changes of molecular species in the liquid silica structure as metallic oxide is added fail to explain the behavior of the liquid silicate? What are the bases of the *network theory* of liquid-silicate structure? Explain why this model fails in describing the glassy state of silicate. Describe the *discreate-polyanion model* of liquid silicates. How does this model explain better the observed experimental facts? What are the disadvantages of this model? Which are the two structures proposed by Tomlinson, White and Bockris to explain the silicate behavior in the composition range from 12-33% M_2O? How do spectroscopic measurements of SiO_4^{-4} lifetime help to understand the liquid silicate structure? What is the type of silicate structure proposed to exist in the earth's interior? Describe the metallurgical process of producing *iron* in a blast furnace. What is *slag*? What is its importance in the manufacture of iron?

5.16 The heat of activation of simple molten salts is generally less than 10 kcal mol^{-1} (42 kJ mol^{-1}), while that of SiO_2 is 59 kcal mol^{-1} (245 kJ mol^{-1}) (cf. Problem 5.32). Similarly, the heat of activation when a metal oxide is

added to liquid silica decreases 2 to 4 times (cf. Fig. 5.71 in the textbook). Explain these significant differences in the activation energies of molten salts, silica, and silica with additives. Which are the most probable rate determining steps in each case? (Cf. Exercise 5.15 in the textbook) (Bockris-GamboaAldeco)

Answer:

Molten salts are constituted by ionic species such as Na^+ and Cl^- in NaCl. Each one of these ions may easily jump into the holes formed in the liquid. Liquid silica, on the other hand, is an associated liquid, constituted by chains of SiO_4 segments. Thus, before jumping, some energy is needed to break one of these segments from the chain, a step not needed in the transport of molten salts. This extra energy is what makes the heat of activation of molten silica larger than that of the molten salt. Now, when a metal oxide such as Na_2O is added to the liquid silica, the Na^+ and O^{-2} ions help in the breaking process of the chain (cf. Problem 5.32), resulting in a decrease of the energy of activation. If the transport process occurs through vacancies, then the most likely rate-determining step (rds) for molten salts such as NaCl is the *formation of voids* in the liquid. For liquid silica, this step is constituted by the *breaking* of the chain structure to give a transport entity. Addition of Na_2O facilitates this step.

5.17 Determine the average radius and the corresponding volume of the hole formed in fused silica at 2100 K. Does a segment of the SiO_4 chain fit in the average hole formed? The surface tension of fused silica is 250 dyn cm^{-1}, and the Si-Si distance is 3.2 Å (cf. Problem 5.23). (GamboaAldeco)

Answer:

The average radius of the holes formed in fused silica is (cf. Eq. 5.44 in the textbook),

$$\langle r \rangle = 0.51 \sqrt{\frac{kT}{\gamma}}$$

$$= 0.51 \sqrt{\frac{\left(1.381 \times 10^{-23} \text{ J K}^{-1}\right)\left(2100 \text{ K}\right)}{250 \text{ dyn cm}^{-1}} \times \frac{10^7 \text{ dyn cm}}{1 \text{ J}}} \qquad (5.48)$$

$$= 1.74 \times 10^{-8} \text{ cm}$$

and the corresponding volume (cf. Eq. 5.144 in the textbook),

$$\langle v_h \rangle = 1.6 \left(\frac{kT}{\gamma} \right)^{3/2}$$

$$= 0.51 \left(\frac{\left(1.381 \times 10^{-23} \text{ JK}^{-1} \right)(2100 \text{ K})}{250 \text{ dyn cm}^{-1}} \times \frac{10^7 \text{ dyn cm}}{1 \text{ J}} \right)^{3/2} \quad (5.49)$$

$$= 2.01 \times 10^{-23} \text{ cm}^3$$

Comment: Consider that the radius of the SiO_4 segment is about half of the Si-Si length. Then, the radius of the SiO_4 segment is 1.6×10^{-8} cm, which fits adequately in the holes formed.

5.18 (a) Assess the total number of individual Si-O bonds in one mole of SiO_2. (b) Give a chemical explanation of why the addition of Na_2O to silicate causes the breaking up of the tetrahedral network. (Cf. Exercise 5.13 in the textbook) (Xu)

Answer:

(a) Since Si is tetra-coordinated, i.e., every Si is surrounded by 4 Si-O bonds, the number of Si-O bonds (N_{Si-O}) in 1.0 mol of SiO_2 is

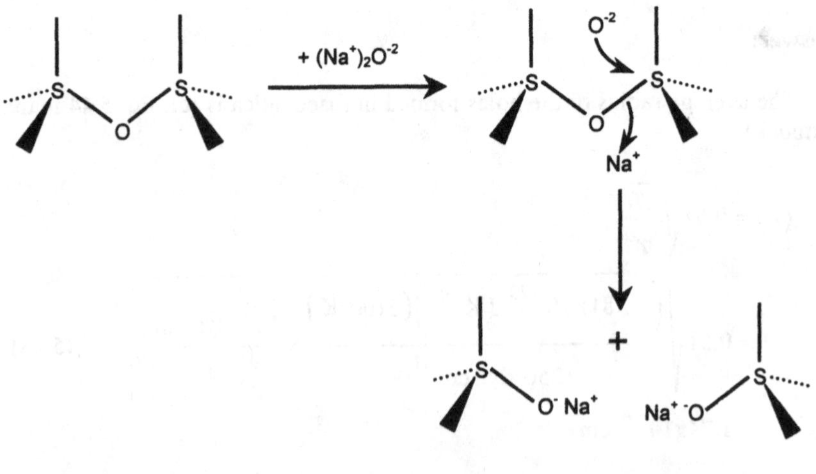

Figure 5.3. Breaking of the SiO_2 network by a Lewis base.

$$N_{Si-O} = 4N_{Si} = 4(1\text{mol}) \times \frac{6.022 \times 10^{23}}{1\text{mol}} = 2.5 \times 10^{24} \text{ bonds} \qquad (5.50)$$

(b) The highly charged Si^{+4} is a strong Lewis acid in the molten state. This Lewis acidic center is easily attacked by the base added, i.e., O^{-2} anions from Na_2O, as shown in Fig. 5.3. The result of the reaction is that the bridging oxygen is replaced by electronically richer oxygen anions, better satisfying the electrophilic Si^{+4} center. The calculation in Problem 5.23 confirms the great repulsion between two neighboring Si^{+4} centers at their equilibrium distance.

5.19 (a) Which is the O/Si ratio of 50% mol CaO in SiO_2? Of 10% mol K_2O? (b)What is the composition of a $M_2O + SiO_2$ mixture that has a O/Si ratio of 3.5? (GamboaAldeco)

Answer:

(a) A 50% mol CaO is written as:

$$1CaO + 1 \ SiO_2 \ \rightarrow \ 3 O : 1Si \ \rightarrow \ \frac{O}{Si} = \frac{3}{1} = 3.$$

Similarly, a 10% mol K_2O is written as:

$$1 K_2O + 9 \ SiO_2 \ \rightarrow \ 19 O : 9 \ Si \ \rightarrow \ \frac{O}{Si} = \frac{19}{9} = 2.1$$

(b) To find the composition of a O/Si = 3.5 ratio, it is needed to find the value of x in the following equation:

$$xM_2O + (100 - x)SiO_2 \ \rightarrow \ \frac{O}{Si} = \frac{x + 2(100 - x)}{100 - x} = 3.5$$
$$\rightarrow \ x = 33.8 \ \rightarrow \ 33.8 \ \% \ M_2O$$

5.20 *A frozen liquid can also flow.* Researchers have found that window glass of many medieval churches in Europe has thicker bottom than top (as large as in the millimiter range). This deformation is evidently caused by the flowing of the silicate. Calculate how far the moving species in the glass can travel in a millennium at room temperature. Does your calculation explain the enlargement in the millmiter range? Which other force should

be taken into account? Consider the glass to have a composition 40 % mol Na_2O, with ring structures of approximatelly 10 Å. At the temperature of glass transition, T_g, the viscosity is $\eta = 10^{12}$ Pa s. (Cf. Exercise 5.14 in the textbook) (Xu)

Data:

40% mol Na_2O	ring structures of 10 Å
$\eta = 10^{12}$ Pa s	$t = 1000$ years

Answer:

At approximatelly 40% Na_2O, the major species that are available for moving is polyanions with ring structures, i.e., $Si_3O_9^{-6}$ (cf. Section 5.13.8 of the textbook). Their sizes are of the order of 10 Å. The mean square distance the ion can travel during this time period is given by the Einstein equation (cf. Eq. 4.27 in the textbook),

$$\langle x \rangle = \sqrt{2D\tau} \qquad (5.51)$$

Assuming that the viscosity of the glass at room temperature is of comparable value to that of the glass-transition temperature, T_g, then, $\eta_{25\ C}$ $=1.0 \times 10^{13}$ Pa s, and the diffusivity of the mobile species is given by (cf. Eq. 5.58 in the textbook),

$$D = \frac{kT}{6\pi\eta r} = \frac{\left(1.381 \times 10^{-23}\ JK^{-1}\right)\left(298\,K\right)}{6\pi\left(1.010^{13}\ Pa\,s\right)\left(10^{-9}\ m\right)} \times \frac{1\,N\,m}{1\,J} \times \frac{1\,Pa}{1\,N\,m^{-2}} \qquad (5.52)$$

$$= 2.18 \times 10^{-26}\ m^2\,s^{-1}$$

A millenium in seconds is:

$$t = 1000\ years \times \frac{365\,days}{1\,year} \times \frac{24\,hr}{1\,day} \times \frac{3600\,s}{1\,hr} = 3.15 \times 10^{10}\ s \qquad (5.53)$$

Substituting the values of D and t into Eq. (5.51) gives,

$$\langle x \rangle = \sqrt{2\left(2.18 \times 10^{-26} \ m^2 s^{-1}\right)\left(3.15 \times 10^{10} \ s\right)} = 3.7 \times 10^{-8} \ m$$

$$\approx 10^{-5} \ mm$$

(5.54)

The above result can be even bigger when the gravitational force is taken into account. In this case the mobile species do not diffuse randomly, i.e., in every direction, but directionally, i.e., downward.

5.21 In the *discrete- polyanion model of liquid silicate*, the entities present vary according to the amount of M_2O added. Draw schematics of the structures present at (a) 0% M_2O, (b) 10% M_2O, (c) 30% M_2O, (d) 50% M_2O, (e) 60% M_2O, and (f) 90% M_2O. What is the Si:O ratio in each case? (g) What is the meaning of an O/Si > 4? (h) How are these entities different from those proposed by the *network theory of liquid silicates*? (Cf. Exercise 5.26 in the textbook) (GamboaAldeco)

Answer:

(a) At 0% M_2O the theory proposes continueous 3-dimentional networks of SiO_4 represented by the drawing in Fig. 5.4.

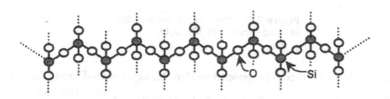

Figure 5.4. Schematic of liquid silicate

At this composition, 100 SiO_2, and $\dfrac{O}{Si} = \dfrac{200}{100} = 2$.

(b) At 10% M_2O the theory proposes SiO_4 network with some broken bonds, as shown in Fig. 5.5. At this composition, 10 M_2O + 90 SiO_2, and $\dfrac{O}{Si} = \dfrac{180 + 10}{90} = 2.1$

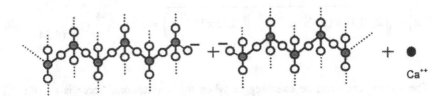

Fig. 5.5 Schematic of the structure of liquid silicate with 10% M_2O

(c) At 30% M_2O the theory proposes $[Si_6O_{15}]^{-6}$ rings, as in Fig. 5.6. At this composition, 30 M_2O + 70 SiO_2, and $\dfrac{O}{Si} = \dfrac{140+30}{70} = 2.4$.

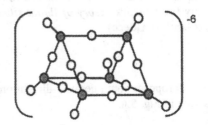

Figure 5.6. Schematic of the structure of liquid silicate with 30% M_2O.

(d) At 50% M_2O the theory proposes a mixture of, say, $[Si_4O_{12}]^{-8}$ and $[Si_8O_{20}]^{-8}$, as in Fig. 5.7. At this composition, 50 M_2O + 50 SiO_2, and $\dfrac{O}{Si} = \dfrac{100+50}{50} = 3$.

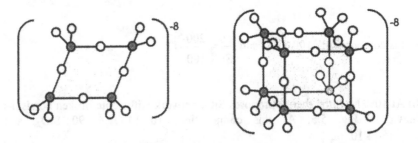

Figure 5.7. Schematic of the structure of liquid silicate with 50% M_2O.

(e) At 60% M_2O the theory proposes chains of $Si_nO_{3n+1}^{-(2n+2)}$, like $[Si_2O_7]^{-6}$ and $[Si_3O_{10}]^{-8}$ as in Fig. 5.8. At this composition, 60 M_2O + 40 SiO_2, and

$$\frac{O}{Si} = \frac{80+60}{40} = 3.5$$

Figure 5.8. Schematic of the structure of liquid silicate with 60% M_2O.

(f) At 90% M_2O the theory proposes SiO_4^{-4} and O^{-2}, as in Fig. 5.9. At this composition, 90 M_2O + 10 SiO_2, and $\dfrac{O}{Si} = \dfrac{20+90}{10} = 11$

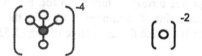

Figure 5.9. Schematic of the structure of liquid silicate with 90% M_2O.

(g) Since the saturated valency of Si is 4, an O/Si > 4 indicates that there is an excess oxygen in the mixture, represented as O^{-2}.

(h) Some of the main differences between the two models are: (i) In the network theory, the cations and the anions are about the same size, in contrast to the discrete-polyanion model, where big ring structures are proposed. (ii) The network theory proposes only breaking of the silicate structure, while the discrete-polyanion model accounts for the breaking of the network and then the formation of different conglomerates.

PROBLEMS

5.22 Using the pair-potentials of one of the pioneers in the modeling of molten salts, i.e., Woodcock and Singer, as well as the corresponding parameter in their work, calculate the equilibrium distance between K^+ and Cl^- ions ($r_{i,j}$) in molten KCl just above the melting point. (Cf. Exercise 5.8 in the textbook) (Bockris-GamboaAldeco)

Answer:

The pairwise addition potentials in molten salts as described by Woodcock and Singer are given by (cf. Eq. 5.10 in the textbook):

$$U_{i,j} = -\frac{z_i z_j e_0^2}{r_{i,j}} + b \exp B\left(\sigma_{i,j} - r_{i,j}\right) + \frac{c_{i,j}}{r_{i,j}^6} + \frac{d_{i,j}}{r_{i,j}^8} \qquad (5.55)$$

The parameters needed for Eq. (5.55) for K^+-Cl^- as used by Woodcock and Singer are given in Table 5.12 in the textbook, i.e.,: $\sigma_{i,j} = 3.048 \times 10^{-8}$ cm, $b = 0.338 \times 10^{-12}$ erg, $B = 2.97 \times 10^8$ cm^{-1}, $c_{i,j} = -48.0 \times 10^{-60}$ erg cm^6, and $d_{i,j} = -73.0 \times 10^{-76}$ erg cm^8. The curve of $U_{i,j}$ against the separation distance, $r_{i,j}$, for opposite charge ions passes through a minimum (cf. Fig. 5.12 in the textbook), which correspond to the point of minimum energy for the system. One way to calculate this minimum is to take the derivative of Eq. (5.55) and make it equal to zero:

$$\frac{z_i z_j e_0^2}{4\pi\varepsilon_0 r_{i,j}^2} - bB \exp B\left(\sigma_{i,j} - r_{i,j}\right) - \frac{c_{i,j}}{r_{i,j}^7} - \frac{d_{i,j}}{r_{i,j}^9} = 0 \qquad (5.56)$$

Arranging terms in Eq. (5.56):

$$\frac{z_i z_j e_0^2}{4\pi\varepsilon_0} r_{i,j}^7 - bB r_{i,j}^9 \exp B\left(\sigma_{i,j} - r_{i,j}\right) - c_{i,j} r_{i,j}^2 - d_{i,j} = 0 \qquad (5.57)$$

Substituting the corresponding parameters,

$$\left(2.307 \times 10^{-19} \text{ erg cm}\right) r_{i,j}^7 \qquad (5.58)$$

$$-\left(1.00 \times 10^{-4} \text{ erg cm}^{-1}\right) r_{i,j}^9 e^{2.97 \times 10^8 \text{ cm}^{-1} \left(3.05 \times 10^{-8} \text{ cm} - r\right)}$$

$$+\left(48 \times 10^{-60} \text{ erg cm}^6\right) r_{i,j}^2 + 73 \times 10^{-76} \text{ erg cm}^8 = 0$$

The substitution of values of $r_{i,j}$ into Eq. (5.58) gives as result the values in the next table, where:

$$A = \left[\left(2.307 \times 10^{-19}\right) r_{i,j}^7\right]$$

$$B = \left[\left(1.00 \times 10^{-4}\right) r_{i,j}^9 \exp\left[2.97 \times 10^8 \left(3.05 \times 10^{-8} - r\right)\right]\right]$$

$$C = \left[\left(48 \times 10^{-60}\right) r_{i,j}^2 + 73 \times 10^{-76}\right]$$

r_{ij} (cm)	A	B	C	l.h.s. Eq. (5.58)
1.0×10^{-8}	0.02×10^{-73}	-0.04×10^{-72}	1.21×10^{-74}	-2.94×10^{-73}
2.0×10^{-8}	2.95×10^{-73}	-1.15×10^{-72}	2.65×10^{-74}	-8.28×10^{-73}
2.3×10^{-8}	7.85×10^{-73}	-1.66×10^{-72}	3.27×10^{-74}	-8.42×10^{-73}
2.4×10^{-8}	10.6×10^{-73}	-1.87×10^{-72}	3.49×10^{-74}	-7.75×10^{-73}
2.5×10^{-8}	14.1×10^{-73}	-1.94×10^{-72}	3.73×10^{-74}	-4.93×10^{-73}
2.6×10^{-8}	18.5×10^{-73}	-2.05×10^{-72}	3.97×10^{-74}	-1.60×10^{-73}
2.7×10^{-8}	24.1×10^{-73}	-2.14×10^{-72}	4.23×10^{-74}	$+3.12 \times 10^{-73}$
2.8×10^{-8}	31.1×10^{-73}	-2.20×10^{-72}	4.49×10^{-74}	$+9.55 \times 10^{-73}$
2.9×10^{-8}	39.8×10^{-73}	-2.25×10^{-72}	4.77×10^{-74}	$+17.77 \times 10^{-73}$
3.0×10^{-8}	50.4×10^{-73}	-2.27×10^{-72}	5.05×10^{-74}	$+28.20 \times 10^{-73}$

A plot of the last column of the above table against the separation distance is given in Fig. 5.10. The l.h.s. of Eq. (5.58) equals zero when $r = 2.62 \times 10^{-8}$ cm, which corresponds to the distance of minimum energy between the K^+ and Cl^- ions.

5.23 A pairwise potential widely used in both Monte Carlo and MD calculations (cf. Eq. 5.10 in the textbook),

$$\Phi_{i,j}(r) = A_{i,j}\frac{z_i z_j}{r_{i,j}} + B_{i,j}\exp C_{i,j}(\sigma_{i,j} - r_{i,j}) + D_{i,j}r_{i,j}^{-6} + E_{i,j}r_{i,j}^{-8}$$

$$(5.59)$$

which describes the potential as function of distance between the two ions i and j. The parameters z_i, z_j are the charges on i and j respectively, while $\sigma_{i,j}$ is the size parameter of the ion pair (normally the sum of the crystallographic radii of i and j). $A_{i,j}$, $B_{i,j}$, $C_{i,j}$, $D_{i,j}$, and $E_{i,j}$ are constants estimated from studies on the crystal of the corresponding salt. (a) Identify the term that dominates the attraction between a pair of opposite charged ions at long range, and the term that prevents these two ions from "falling into each other." (b) Which is the parameter in the second term that determines the "steepness" of the repulsion felt by these two ions once their size parameters and center-to-center distances have been fixed? (c) In molten silicate, the Si-Si equilibrium distance is ca. 3.2 Å. Determine by calculation whether the force due to the second term or the Coulombic like charge repulsion dominates. What does the result connive at concerning the stability of the silicate network? Consider: $A_{i,j} = e_0^2$, $B_{i,j} = 0.19 \times 10^{-19}$ J, $C_{i,j} = 3.44 \times 10^{10}$ m^{-1}, and $r_{crys,Si} = 1.31 \times 10^{-10}$ m. (Cf. Problem 5.18 in the textbook) (Xu)

Answer:

(a) The terms in the equation are in turn: Coulombic term, nuclear repulsion term, dipole-dipole attraction term, and ion-dipole attraction term. The last two

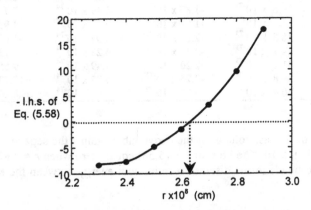

Figure 5.10. Determination of the equilibrium distance between K$^+$ and Cl$^-$ ions in molten KCl according to Woodcock and Singer theory.

terms contain r^{-6} and r^{-8}, and thus, can only exert influence in a very short range. In the long range, only the first Coulombic term works as major source of attraction, i.e., between opposite charged particles. Thus, for the case where $z_i z_j$ < 0, the contribution of this term is negative, since the attraction lowers the potential of the system. The second term, the nuclear repulsion term, prevents these two opposite charge ions from "falling into each other." As r decreases, the positive contribution of this term rises exponentially, making the potential of the system higher than zero (no interaction state).

(b) The parameter that accounts for the repulsive interaction is C_{ij}, known as the "softness parameter". Considering the repulsive term in Eq. (5.59),

$$\Phi_{i,j}\left(rep\right) = B_{i,j}\; exp\, C_{i,j}\left(\sigma_{i,j} - r_{i,j}\right) \tag{5.60}$$

Taking logarithms on both sides of this equation,

$$ln\, \Phi_{i,j}\left(rep\right) = ln\, B_{i,j} + C_{i,j}\sigma_{i,j} - C_{i,j}r_{i,j} \tag{5.61}$$

The slope (or steepness) of the curve $ln\, \Phi_{i,j}\left(rep\right)$ against r is given by

$$\frac{\partial \Phi_{i,j}\left(rep\right)}{\partial r} = -C_{i,j} \tag{5.62}$$

Putting it in words, when two ions approach each other, the increase of the repulsion potential depends on the value of $C_{i,j}$.

(c) For molten silicate, at the equilibrium distance of 3.2×10^{-10} m the two last terms in Eq. (5.59) can be considered negligible. Thus,

$$\Phi_{i,j}\left(r\right) = A_{i,j}\frac{z_i z_j}{r_{i,j}} + B_{i,j}\; exp\left[C_{i,j}\left(\sigma_{i,j} - r_{i,j}\right)\right] \tag{5.63}$$

$$= \Phi_{i,j}\left(Coul\right) + \Phi_{i,j}\left(rep\right)$$

Since force is defined as (cf. Table 4.1 in the textbook) $Force = -\dfrac{\partial \Phi}{\partial r}$, then

$$F\left(Coul\right) = -\frac{\partial \Phi_{i,j}\left(Coul\right)}{\partial r} = A_{i,j}\frac{z_i z_j}{r_{i,j}^2} \tag{5.64}$$

and

$$F(rep) = -\frac{\partial \Phi_{i,j}(rep)}{\partial r} = B_{i,j}C_{i,j}\,exp\left[C_{i,j}\left(\sigma_{i,j} - r_{i,j}\right)\right] \quad (5.65)$$

At $r = 3.2 \times 10^{-10}$ m, Eq. (5.64) and (5.65) become, in the *mksa* system,

$$F(Coul) = \frac{z_{Si}^2 e_0^2}{4\pi\varepsilon_0 r^2} = \frac{(4)^2\left(1.602 \times 10^{-19}\ C\right)^2}{\left(1.112 \times 10^{-10}\ C^2 J^{-1} m^{-1}\right)\left(3.2 \times 10^{-10}\ m\right)^2} \quad (5.66)$$

$$= 3.604 \times 10^{-8}\ N$$

and

$$F(rep) = \left(0.19 \times 10^{-19}\ J\right)\left(3.44 \times 10^{10}\ m^{-1}\right)$$
$$\times exp\left\{3.44 \times 10^{10}\ m^{-1}\left[3.2 - 2(1.31)\right] \times 10^{-10}\ m\right\} = 0.481 \times 10^{-8}\ N \quad (5.67)$$

At the equilibrium distance of 3.2 Å, the contribution to the repulsive force is mainly from Coulombic like charge. The above results show the great repulsion existing between Si^{+4} centers, and therefore, the possibility of breaking up the network when a nucleophile is present (cf. Exercise 5.18).

5.24 In the Fürth-hole model for molten salts, the primary attraction is that it allows a rationalization of the empirical expression $E^{\neq} = 3.74\ RT_{m.p.}$. In this model, fluctuations of the structure allow openings (holes) to occur and to exist for a short time. (a) Using the probability curve obtained for NaCl at 900 ^0C in Exercise 5.3, probe that the mean-hole size turns out to be about the size of the ions in the molten salt. (b) Determine the probability of finding a hole that is two times the most probable hole size, and (c) the probability of finding a hole that would allow paired-vacancy diffusion. Comment on your results. (Cf. Problem 5.13 in the textbook) (Bockris-GamboaAldeco)

Answer:

(a) The most probable hole size is given by the maximum in the plot P_r vs. r. Thus, from Exercise 5.3, this graph was obtained for NaCl, with a maximum at a hole radius of **190 pm**. Now, the area under the whole curve P_r vs. r represents the total probability of finding holes any size, i.e.,

$$\int_0^\infty P_r\, dr = 1 \qquad (5.68)$$

The fraction of holes of sizes "around" the most probable (or must populous) radius r_{max} is represented as $r_{max} \pm \Delta r$, and is given by

$$\frac{\int_{r_{min}+\Delta r}^{r_{max}+\Delta r} P_{r_{max}}\, dr}{\int_0^\infty P_r\, dr} = \int_{r_{min}+\Delta r}^{r_{max}+\Delta r} P_{r_{max}}\, dr \qquad (5.69)$$

The next step is to decided what the value of Δr is. To keep the hole size close to the value of r_{max}, the variation is allowed to be, say 1% of r_{max}. Thus, $\Delta r = 0.01 r_{max} = 0.01(190 \text{ pm}) \approx 2$ pm, that is, 188 pm $< r_{max} <$ 192 pm. To evaluate the integral in Eq. (5.69), one makes use of the P_r vs. r plot in Fig. 5.1. At $r_{max} = 190$ pm, $P_{r_{max}} = 0.007375\,\text{pm}^{-1}$. Then the integral in Eq. (5.69) can be approximated as

$$\int_{188\text{pm}}^{192\text{pm}} P_{r_{max}}\, dr \approx P_{r_{max}} (2\Delta r) = \left(0.007375\,\text{pm}^{-1}\right)(4\text{pm}) = 0.028 \qquad (5.70)$$

That is, about 3% of the holes have a radius between 192 pm and 188 pm. Since $r_{Na^+} = 95$ pm and $r_{Cl^-} = 181$ pm, only *one* ion is allowed to jump into one hole.

(b) The probability of finding a hole twice as large as the most probable hole size is equivalent to finding the probability of the existence of a hole of radius equal to $r_{hole} = 2r_{max} = 2(190 \text{ pm}) = 380$ pm. At this value of r, an $P_{r_{hole}} = 0.000058$ pm^{-1} is obtained from Exercise 5.3. With $\Delta r = 2$ pm,

$$\int_{378\text{pm}}^{382\text{pm}} P_{r_{hole}}\, dr \approx P_{r_{hole}} (2\Delta r) = \left(0.000058\,\text{pm}^{-1}\right)(4\text{pm}) = 0.00023 \qquad (5.71)$$

that is, 0.023% of the holes have a radius between 378 and 382 pm, which corresponds only to approximately 1% of the probability of finding a hole of radius r_{max}.

(c) To fit a pair Na^+-Cl^- into a hole, one has to sum their radii, and consider the decrease of the internuclear distance in the fused salt as compared to that in the crystal. The internuclear distance between the two ions is $r_{Na^+} + r_{Cl^-} = 95\,pm + 181\,pm = 276\,pm$. The internuclear distance in fused salts shrinks by approximately 5% (cf. Table 5.9 in the textbook). Thus, $r_{pair} = 0.95(276\,pm) = 262\,pm$. At this radius, $P_{r_{pair}} = 0.0034\,pm^{-1}$ is obtained from Exercise 5.3. With $\Delta r = 2$ pm,

$$\int_{P_{r_{pair}} - \Delta r}^{P_{r_{pair}} + \Delta r} P_{r_{hole}}\, dr \approx P_{r_{pair}} \left(2\Delta r\right) = \left(0.0034\,pm^{-1}\right)(4\,pm) = 0.014 \qquad (5.72)$$

i.e., 1.4% is the probability of finding a hole of the size of a Na^+-Cl^- pair. This corresponds to about half the probability of finding a hole of r_{max}.

5.25 (a) A general equation for obtaining average values is $\langle x \rangle = \int_0^\infty x P_x\, dx$.

This equation, together with the results of the hole model, was used in the textbook to obtain the average radius of a hole for ionic liquids. Using a similar procedure, derive the average surface area, $\langle s \rangle$, of the holes in ionic liquids. (b) Determine again the average surface area of the holes, $\langle s' \rangle$, , but using this time the equation of the average radius of a hole. What conclusions can you withdraw from comparing the surface areas $\langle s \rangle$, and $\langle s' \rangle$? (c) Do similar calculations to determine the average volume of the holes, i.e., $\langle v \rangle$ and $\langle v' \rangle$. Hint: The gamma function, defined as $\Gamma(n) = \int_0^\infty t^{n-1} e^{-t}\, dt$ has the following properties: (i) $\Gamma(1/2) = \sqrt{\pi}$, (ii) $\Gamma(n+1) = n!$, when n is a positive integral, (iii) $\Gamma(n+1) = n\,\Gamma(n)$ when n is a positive real. (Cf. Problem 5.3 in the textbook) (Xu)

Answer:

(a) The holes in ionic liquids can be viewed as spheres with radius r, and thus, with surface area $s = 4\pi r^2$. The average surface area can be obtained from the general procedure of obtaining average properties with known probability (cf. Eq. 5.40 in the textbook). In this case, the probability of the existence of a given hole is given by (cf. Eq. 5.39 in the textbook)

$$P_r\, dr = \frac{16}{15\pi^{1/2}} a^{7/2} r^6 e^{-ar^2}\, dr \qquad (5.73)$$

with $a = 4\pi\gamma / dT$ (cf. Eq. 5.35 in the textbook). Therefore, the average-surface area $\langle s \rangle$ is,

$$\langle s \rangle = \int_0^\infty s \cdot P_r \, dr = \int_0^\infty \left(4\pi r^2 \, \frac{16}{15\pi^{1/2}} a^{7/2} r^6 e^{-ar^2} \right) dr$$

$$= \frac{64\sqrt{\pi}}{15} a^{7/2} \int_0^\infty r^8 e^{-ar^2} \, dr \tag{5.74}$$

If $t = ar^2$, then $r = (t/a)^{1/2}$ and $dr = 1/2 (at)^{1/2} \, dt$. Therefore, Eq. (5.74) becomes,

$$\langle s \rangle = \frac{32}{15} \frac{\pi^{1/2}}{a} \int_0^\infty t^{7/2} e^{-t} \, dt \tag{5.75}$$

The integral $\int_0^\infty t^{7/2} e^{-t} \, dt$ is the gamma function, $\Gamma(n) = \int_0^\infty t^{n-1} e^{-t} \, dt$, when $n = 9/2$ (cf. Appendix 5.2 in the textbook). Thus,

$$\int_0^\infty t^{7/2} e^{-t} \, dt = \Gamma\left(\frac{9}{2}\right) \tag{5.76}$$

and Eq. (5.75) becomes,

$$\langle s \rangle = \frac{32}{15} \frac{\pi^{1/2}}{a} \Gamma\left(\frac{9}{2}\right) \tag{5.77}$$

To evaluate the gamma function we make use of its properties. Thus,

$$\Gamma\left(\frac{9}{2}\right) = \frac{7}{2}\Gamma\left(\frac{7}{2}\right) = \frac{7}{2} \times \frac{5}{2}\Gamma\left(\frac{5}{2}\right) = \frac{7}{2} \times \frac{5}{2} \times \frac{3}{2}\Gamma\left(\frac{3}{2}\right)$$

$$= \frac{7}{2} \times \frac{5}{2} \times \frac{3}{2} \times \frac{1}{2}\Gamma\left(\frac{1}{2}\right) = \frac{7}{2} \times \frac{5}{2} \times \frac{3}{2} \times \frac{1}{2} \times \sqrt{\pi} = \frac{105}{16}\sqrt{\pi} \tag{5.78}$$

Substituting Eq. (5.78) and the constant $a = 4\pi\gamma / dT$ into Eq. (5.77),

$$\langle s \rangle = \frac{7}{2} \frac{kT}{\gamma} = 3.5 \frac{kT}{\gamma} \tag{5.79}$$

(b) The average-hole radius, $\langle r \rangle$, is given by (cf. Eq. 5.44 in the textbook),

$$\langle r \rangle = \frac{8}{5\pi} \sqrt{\frac{kT}{\gamma}} \tag{5.80}$$

If it is considered that all the holes are the same size, then the surface area of the holes, s', is given by,

$$\langle s' \rangle = 4\pi \langle r \rangle^2 = 4\pi \left(\frac{8}{5\pi} \sqrt{\frac{kT}{\gamma}} \right)^2 = \frac{256}{25\pi} \frac{kT}{\gamma} = 3.3 \frac{kT}{\gamma} \tag{5.81}$$

which is in good agreement with the average surface area calculated in (a). This means that all the holes are approximately the same size, i.e., the average size.

(c) Similarly to Eq. (5.74), the average-hole volume can be calculated as

$$\langle v \rangle = \int_0^\infty v \cdot P_r \, dr = \int_0^\infty \left(\frac{4}{3} \pi r^3 \cdot \frac{16}{15\pi^{1/2}} a^{7/2} r^6 e^{-ar^2} \right) dr$$
$$= \frac{64\sqrt{\pi}}{45} a^{7/2} \int_0^\infty r^9 e^{-ar^2} \, dr \tag{5.82}$$

If $t = ar^2$, $\langle v \rangle$ becomes,

$$\langle v \rangle = \frac{32}{45} \frac{\pi^{1/2}}{a^{3/2}} \int_0^\infty t^4 e^{-t} \, dt = \frac{32}{45} \frac{\pi^{1/2}}{a^{3/2}} \Gamma(5) \tag{5.83}$$

However, $\Gamma(5) = 4! = 24$. Thus,

$$\langle v \rangle = \frac{256}{15} \frac{\pi^{1/2}}{a^{3/2}} = \frac{256}{15} \pi^{1/2} \times \frac{1}{8\pi^{3/2}} \sqrt{\left(\frac{kT}{\gamma} \right)^3} = \frac{32}{15\pi} \sqrt{\left(\frac{kT}{\gamma} \right)^3} \tag{5.84}$$

$$= 0.68 \sqrt{\left(\frac{kT}{\gamma}\right)^3}$$

Now, considering the average-hole radius, $\langle r \rangle$,

$$\langle v' \rangle = \frac{4}{3}\pi \langle r \rangle^3 = \frac{4}{3}\pi \left(\frac{8}{5\pi}\sqrt{\frac{kT}{\gamma}}\right)^3 = \frac{2048}{375\pi^2}\sqrt{\left(\frac{kT}{\gamma}\right)^3} \qquad (5.85)$$

$$= 0.55 \sqrt{\left(\frac{kT}{\gamma}\right)^3}$$

which supports the result in **(b)**.

5.26 Using the results of Problem 5.25, calculate the work needed to make a hole of average size at 900 ^{0}C in any molten salt if the Fürth's "nearly-boiling assumption" holds. (Cf. Problem 5.3 in the textbook) (Xu)

Answer:

With Fürth's assumption that molten salts are at nearly boiling state (i.e., the internal and external pressure experienced by the hole balance each other), the work in forming a hole comes only from the surface work contribution (cf. Eq. 5.33 in the textbook);

$$W = \langle s \rangle \gamma \qquad (5.86)$$

The average surface area of holes in ionic liquids is given by the results obtained in Eq. (5.79) in Problem 5.25(a), namely,

$$\langle s \rangle = \frac{7}{2}\frac{kT}{\gamma} \qquad (5.87)$$

Substituting $\langle s \rangle$ from Eq. (5.87) into Eq. (5.86),

$$W = \frac{7}{2}kT = \frac{7}{2}\left(1.38 \times 10^{-23} \text{ JK}^{-1}\right)\left(1173\,\text{K}\right) = 5.66 \times 10^{-20} \text{ J} \qquad (5.88)$$

This is the work needed to form a single hole. For one mole of holes,

$$W = \left(5.66 \times 10^{-20} \text{ J}\right)\left(6.022 \times 10^{23} \text{ mol}^{-1}\right) = 3.41 \times 10^4 \text{ J mol}^{-1} \quad (5.89)$$

5.27 When a liquid supercools (i.e., does not crystallize when its temperature drops below the thermodynamic melting point), the liquid-like structure is frozen due to the high viscosity of the system. The state of supercooled liquid is in a so-called *visco-elastic state*. If the crystallization could be further avoided as temperature continues to drop, *glass transition* happens at a certain temperature, T_g, where the "frozen liquid" turns into a brittle, rigid state known as *glassy state*. A well accepted definition for glass transition indicates its formation when the relaxation time of the system, τ, is 200 s, or its viscosity, η, is 10^{12} Pa s (an arbitrary standard, of course!). (a) Calculate the average distance an ion can travel during the period of a single relaxation time in a substance with room-temperature glass transition. (b) A simple relation between relaxation time and viscosity exists in all liquids down to the glass transition temperature, $\tau = K\eta$, where K has a very small temperature dependence and can be regarded as a constant independent of temperature. Obtain this constant and calculate the theoretical upper limit of the viscosity of liquids, considering the fact that the electronic relaxation time measured in the far-infrared region is 10^{-13} s. (Cf. Problem 5.19 in the textbook) (Xu)

Answer:

(a) The mean square distance the ion can travel during this time period is given by the Einstein equation (cf. Eq. 4.27 in the textbook):

$$\langle x \rangle^2 = 2D\tau \quad (5.90)$$

where τ represents a single relaxation time ($\tau = 200$ s) by the definition of the glass transition temperature (T_g) given above. Assuming the ion's radius is of the order of 10^{-10} m, a typical molecular size, then, its diffusivity is given by the Einstein-Stokes equation (cf. Eq. 5.58 in the textbook):

$$D = \frac{kT}{6\pi\eta r} = \frac{\left(1.381 \times 10^{-23} \text{ JK}^{-1}\right)(298\text{K})}{6\left(10^{12} \text{ Pa s}\right)\left(10^{-10} \text{ m}\right)} \times \frac{1\text{N m}}{1\text{J}} \times \frac{1\text{Pa}}{1\text{N m}^{-2}} \quad (5.91)$$

$$= 2.18 \times 10^{-22} \ m^2 s^{-1}$$

Substituting this value of D into equation (5.90),

$$\langle x \rangle^2 = 2 \left(2.18 \times 10^{-24} \ m^2 s^{-1} \right) (200 \ s) = 8.73 \times 10^{-22} \ m^2 \qquad (5.92)$$

or,

$$\langle x \rangle = \sqrt{8.73 \times 10^{-22} \ m^2} = 2.96 \times 10^{-11} \ m \qquad (5.93)$$

The value of $\langle x \rangle$ from Eq. (5.93) is smaller than the size of the ion.

(b) At the glass transition temperature, T_g, $\tau = 200$ s and $\eta = 10^{12}$ Pa s, and the constant K can be obtained from the given equation $\tau = K\eta$. Thus,

$$K = \frac{\tau}{\eta} = \frac{200 s}{10^{12} \ Pa \ s} = 2 \times 10^{-10} \ Pa^{-1} \qquad (5.94)$$

For any liquid, as temperature goes up the viscosity drops, and the relaxation time becomes shorter and shorter (cf. Eq. 5.94). However, the time cannot become shorter than the electronic relaxation time, in this case, 10^{-13} s. This value is therefore taken as the theoretical upper limit for relaxation time at "extremely-high" temperatures (of course, no liquid can be heated up to this temperature without chemical decomposition, hence it represents a *theoretical* state). At this "extremely-high" temperature, the viscosity is,

$$\eta = \frac{\tau}{K} = \frac{1 \times 10^{-13} \ s}{2 \times 10^{-10} \ Pa^{-1}} = 5.0 \times 10^{-4} \ Pa \ s \qquad (5.95)$$

This is the limiting value of viscosity of liquids at high temperatures.

5.28 Using the data in Table 5.18 in the textbook, determine the self-diffusion coefficient of the given ions at 1100 K. What phenomenological conclusions can you withdraw from these results? Give reasonable explanations for your conclusions. (GamboaAldeco)

Answer:

The table below shows the values of D_0 and $E_D^{\neq}$ for the diffusion coefficient equation given in Table 5.18 in the textbook. The diffusion coefficient can be calculated using these parameters by the equation (cf. Eq. 5.55 in the textbook),

$$D = D_0 \ exp\left(-\frac{E_D^{\neq}}{RT}\right) \tag{5.96}$$

The values of D calculated using Eq. (5.90) are shown in the last column of the following table:

Molten salt/Tracer	$D_0 \times 10^5$ $(cm^2 \ s^{-1})$	$E_D^{\neq}$ $(kJ \ mol^{-1})$	$D \times 10^5$ $(cm^2 \ s^{-1})$
$NaCl/^{22}Na$	2.1	29.87	8.01
$NaCl/^{36}Cl$	1.9	31.09	6.34
$KCl/^{42}K$	1.8	28.78	7.73
$KCl/^{36}Cl$	1.8	29.83	6.89
$CaCl_2/^{45}Ca$	0.38	25.65	2.30
$CaCl_2/^{36}Cl$	1.9	37.07	3.30
$SrCl_2/^{89}Sr$	0.21	22.51	1.79
$SrCl_2/^{36}Cl$	0.77	28.79	3.30
$BaCl_2/^{140}Ba$	0.64	37.49	1.06
$BaCl_2/^{36}Cl$	2.0	39.66	2.62
$CdCl_2/^{115}Cd$	1.1	28.62	4.81
$CaCl_2/^{36}Cl$	1.1	28.45	4.90

Conclusions: (*i*) The monovalent cations have larger D's than the divalent cations (e.g., $D_{Na/NaCl}$ and $D_{Ca/CaCl2}$). (*ii*) As the size of the cation in the same group in the periodic table increases, the diffusion coefficient decreases (e.g., $D_{Ca\ CaCl2}$ and $D_{Ba/BaCl2}$). (*iii*) The diffusion coefficient of the anion in the 1:1 salts is smaller than that of the cation (e.g., $D_{Na\ NaCl}$ and $D_{Cl\ NaClCl}$). (*iv*) The trend in (*iii*) reverses in the 1:2 salts (e.g., $D_{Ca\ CaCl2}$ and $D_{Cl\ CaCl2}$). (*v*) The diffusion coefficient of the anions in the 1:1 salts is twice as large as that in the 1:2 salts (e.g., $D_{Cl/NaCl}$ and $D_{Cl/CaCl2}$).

Explanations: Comparing monovalent and divalent cations, the divalent cations have more bonds to break before they can jump to another site. Their movement through the liquid is therefore more difficult than a monovalent ion with less bonds, explaining their smaller D's [c.f. (*i*) above]. Larger ions need greater local rearrangement at a site before they can jump into it. As a result, the total energy involved in the moving process is larger and diffusion more difficult (i.e., smaller D.) This accounts for the decrease of D as the cations size increases [cf. (*ii*) above], and for the decrease of anion's D as compared with the cation in the 1:1 salts [cf. (*iii*) above]. The smaller D's of the divalent cations as compared to that of the anions can be explained considering that the difference of energy involved in the breaking of bonds is larger than that accounting for the different size of the ions [cf. (*iv*) above]. The jump barrier energy of the chloride ions is larger in the 1:2 salts than in the 1:1 salts. This is due to the larger attraction these ions have with the divalent cations, making their movement more difficult and thus, decreasing their D's [cf. (*v*) above].

5.29 The conductance calculated from the Nernst-Einstein equation is several tenths of percent more than that measured. An interpretation is that the diffusion coefficient includes contributions from jumps into paired vacancies and these (having no net charge) would contribute nothing to the conductance while fully counting for the diffusion. (a) Using the Stokes-Einstein equation, calculate the self-diffusion coefficient of the Na^+ and Cl^- ions and that of the paired Na^+-Cl^- at 1173 K. (b) Calculate the *real* equivalent conductivity, that is, the measured value whether pairs would not be present. (c) Considering now the diffusion due to Na-Cl pairs, find the equivalent conductivity that would have been measured. How much grater would the Nernst-Einstein equation indicate the conductivity to be than it really is? Consider the mean-hole radius of the pair Na^+-Cl^- as 262 pm and those of Na^+ and Cl^- as 95 pm and 181 pm, respectively (cf. Problem 5.22). The viscosity of fused NaCl is 1.05 cpoise. (Cf. Problem 5.10 in the textbook) (Bockris-GamboaAldeco)

Answer:

(a) From Stokes-Einstein equation (cf. Eq. 5.58 in the textbook),

$$D_{pair} = \frac{kT}{6\pi r_{pair}\eta} = \frac{\left(1.38\times10^{-16}\ \text{erg K}^{-1}\right)(1173\,\text{K})}{6\pi\left(262\times10^{-10}\ \text{cm}\right)(0.0105\,\text{poise})}$$

$$\times\frac{1\,\text{poise}}{1\,\text{g cm}^{-1}\text{s}^{-1}}\times\frac{1\,\text{g cm}^2\text{s}^{-2}}{1\,\text{erg}} = 3.12\text{x}10^{-5}\ \text{cm}^2\text{s}^{-1} \qquad (5.97)$$

In the same way, $D_{Na^+} = 8.61 \times 10^{-5}$ cm^2 s^{-1} and $D_{Cl^-} = 4.52 \times 10^{-5}$ cm^2 s^{-1}. These two values were not obtained experimentally, and thus, correspond to individual values of diffusion coefficients, i.e., $D_{Na+,ind}$ and $D_{Cl-,ind}$.

(b) Using the Nernst-Einstein equation for a 1:1 electrolyte (cf. Eq. 5.61 in the textbook):

$$\Lambda' = \frac{F^2}{RT}\left(D_{Na^+,ind} + D_{Cl^-,ind}\right)$$

$$= \frac{\left(96500\,C\,eq^{-1}\right)^2}{\left(8.314\,J\,mol^{-1}\,K^{-1}\right)\left(1173K\right)}(8.61+4.52)\times 10^{-5}\ cm^2\,s^{-1} \quad (5.98)$$

$$\times \frac{1J}{1CV}\times\frac{1V\Omega^{-1}}{1Cs^{-1}} = 125\ S\,cm^2\,eq^{-1}$$

(c) If the diffusion coefficients were determined by tracer experiments instead of calculated as done in (a), then, the equivalent conductivity would have been (cf. Eq. 5.76 in the textbook)

$$\Lambda_{calc} = \Lambda' + \frac{2F^2}{RT}D_{pair}$$

$$= 125S\,cm^2\,eq^{-1} + \frac{2\left(96500\,C\,eq^{-1}\right)^2\left(3.12\times 10^{-5}\ cm^2\,s^{-1}\right)}{\left(8.314\,J\,mol^{-1}\,K^{-1}\right)\left(1173K\right)} \quad (5.99)$$

$$\times \frac{1J}{1CV}\times\frac{1V\Omega^{-1}}{1Cs^{-1}} = 184S\,cm^2\,eq^{-1}$$

Comparing Eqs. (5.98) and (5.99) indicates that the Nernst-Einstein equation would have given a value of the equivalent conductivity 47% larger than it is in reality.

5.30 The relaxation time τ and the viscosity η are variables whose temperature dependence follows a non-Arrhenius behavior, i.e., they are of the type $exp(BT_0/T-T_0)$, instead of the well known Arrhenius trend, i.e.,

$exp(E_a/RT)$. In this non-Arrhenius equation, T_0 represents a characteristic temperature, and the constant B is an important characteristic of the structure of the liquid, whose inverse is known as the *fragility* of the liquid. Thus, the larger the value of B, the *stronger* (or the less fragile) the liquid is. (a) Write the non-Arrhenius equations for relaxation time and viscosity. (b) What type of plots would differentiate the Arrhenius from the non-Arrhenius behavior? (c) Explain how the value of B differentiates these two types of behavior. (d) How would the strong and fragile liquids be identified in the most common plot of *ln* η vs. *1/T*? (Cf. Problem 5.20 in the textbook) (Xu)

Answer:

(a) The non-Arrhenius equations for the relaxation time and the viscosity are:

$$\eta = A \, exp\frac{BT_0}{T-T_0} \quad \text{and} \quad \tau = A' \, exp\frac{BT_0}{T-T_0} . \tag{5.100}$$

(b) To analyze the effect of B on the temperature-dependence of the variable, we can compare an ideal Arrhenius relation with the case of an non-Arrhenius equation. Taking the viscosity as an example, if its temperature dependence is strictly Arrhenius, then,

$$\eta = A \, exp\frac{E_\eta}{RT} \tag{5.101}$$

where E_η is independent of temperature. Then,

$$ln\eta = ln A + \frac{E_\eta}{RT} \tag{5.102}$$

or,

$$\frac{1}{ln\eta - ln A} = \frac{RT}{E_\eta} = KT \tag{5.103}$$

where K is a constant independent of temperature. Therefore, a plot of $1/(ln\,\eta - ln\,A)$ vs. T should give a straight line passing through the origin if the relation between η and temperature is of the Arrhenius type (Fig. 5.11). Similarly, if the temperature dependence is non-Arrhenius,

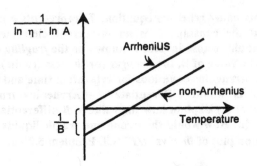

Figure 5.11. Arrhenius and non-Arrhenius behavior of viscosity.

$$\eta = A \, exp \, B \frac{T_0}{T - T_0} \qquad (5.104)$$

or,

$$ln\eta = ln A + \frac{BT_0}{T - T_0} \qquad (5.105)$$

Therefore,

$$\frac{1}{ln\eta - ln A} = \frac{T}{BT_0} - \frac{1}{B} = KT - \frac{1}{B} \qquad (5.106)$$

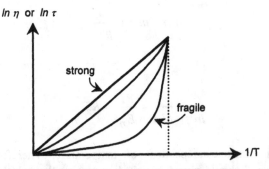

Figure 5.12. Behavior of *strong* and *fragile* liquids in a *ln η* or *ln τ* vs. 1/T graph.

A plot of $1/(\ln \eta - \ln A)$ vs. temperature should be a straight line with a negative intercept of $1/B$ if the relation between η and T is of the non-Arrhenius type (Fig. 5.11).

(c) For a series of similar systems, the larger B is, the smaller the intercept is, and the closer the temperature dependence is to the Arrhenius behavior. Therefore, *strong* liquids (larger B) have more Arrhenius-like temperature-dependence of both, viscosity and relaxation time. On the other hand, the *fragil* liquids have a more canonical temperature dependence of both, viscosity and relaxation time (see Fig. 5.12).

(d) In a $\ln \eta$ or $\ln \tau$ vs $1/T$ plot, the strong and fragil liquids are viewed as represented in Fig. 5.12.

5.31 The Addam-Gibbs theory assumes that the energy barrier to the formation of a transport process is proportional to the product of the activation free energy, $\Delta\mu$, and a certain critical size, z^*, of a subsystem, whose cooperation is needed in the transport. Hence, according to the universal Boltzmann relation, the probability for such a process to happen is,

$$P = exp\left[-\frac{\Delta\mu S_c^* T_0}{kC(T-T_0)} \right] \qquad (5.107)$$

where $\Delta\mu$, S_c^* and T_0 could be viewed as constants independent of temperature but characteristic of the system. Prove that the temperature-dependencies of the relaxation time τ and viscosity η of super-cooled liquids are both non-Arrhenius, i.e., of the $exp(BT_0/T-T_0)$ type, where T_0 is a characteristic temperature. (Cf. Exercise 5.29 in the textbook) (Xu)

Answer:

The viscosity, which is resistance to the transport process, should be inversely proportional to the probability of the process to happen with a pre-exponential term A. Using Eq. (5.107), the viscosity can be written as,

$$\eta = \frac{A}{P} = A\,exp\left[\frac{\Delta\mu S_c^* T_0}{kC(T-T_0)} \right] = A\,exp\left[\frac{BT_0}{T-T_0} \right] \qquad (5.108)$$

where B is a constant independent of temperature, i.e., $B = \Delta\mu\, S_c^*/kC$. The pre-exponential term, A, may or may not have temperature dependence, but its influence on the temperature dependence is trivial as long as $T >> T_0$, i.e., in the state of most *supercooled liquids* and *all true liquids*. Thus, from $\tau = K\eta$, a similar temperature dependence can be derived, for the relaxation time, i.e.,

$$\tau = A'\exp\left[\frac{BT_0}{T-T_0}\right] \qquad (5.109)$$

5.32 $Ca(NO_3)_2$-KNO_3 or better known as CKN, is a well known molten salt that easily vitrifies upon cooling. An attempt to ascertain the fragility of this system (cf. Problem 5.30) was made on a CKN sample with a glass temperature of T_g of 80 ^0C (cf. Problem 5.27). This sample was heated up to 120 ^0C and its dielectric-relaxation time was measured by an impedance bridge as 10^{-2} s. Using also the data from Problems 5.27 and 5.30 classify this ionic liquid. (Cf. Exercise 5.30 in the textbook) (Xu)

Answer:

Using the equation Eq. (5.100),

$$\tau = A'\exp\frac{BT_0}{T-T_0} \qquad (5.110)$$

When $T = \infty$, the exponential in Eq. (5.110) becomes $exp(BT_0/\,T\text{-}T_0) \rightarrow 1$, and therefore, $\tau = A'$. Thus, A' becomes the value of the relaxation time at extremely high temperature. From Problem 5.27 it is known that $A' = 10^{-13}$ s. At $T = T_g = 80\ ^0C = 353$ K, the relaxation time is $\tau = 200$ s, by definition of T_g. Hence, from Eq. (5.110)

$$200\,s = \left(10^{-13}\,s\right)\exp\frac{BT_0}{353-T_0} \qquad (5.111)$$

In the same way, at $T = 120\ ^0C = 393$ K, the relaxation time is $\tau = 0.01$ s,

$$0.01\ s = \left(10^{-13}\,s\right)\exp\frac{BT_0}{393-T_0} \qquad (5.112)$$

From these two equations, the values of B and T_0 can be obtained. Thus, from Eq. (5.111),

$$ln\,200 = ln\,10^{-13} + \frac{BT_0}{353 - T_0} \qquad \text{or} \qquad BT_0 = 35.2\,(353 - T_0) \quad (5.113)$$

and from Eq. (5.112),

$$ln\,0.01 = ln\,10^{-13} + \frac{35.2\,(353 - T_0)}{353 - T_0} \qquad\qquad (5.114)$$

or $T_0 = 250$ K and $B = 14.4$.   The value of $B = 14.4$ is a small value compared with 100 for pure silicate (cf. Problem 5.27). Therefore, CKN is a rather fragile liquid.

5.33 Calculate the transport numbers of the cation and the anion in molten CsCl at 943 K. The experimental equivalent conductivity of the fused salt is 67.7 S cm^2 eq^{-1}.. The observed diffusion coefficients of Cs$^+$ and Cl$^-$ ions in molten CsCl are 3.5x10^{-5} cm^2 s^{-1} and 3.8x10^{-5} cm^2 s^{-1}, respectively. (Cf. Problem 5.18 in the textbook) (Contractor)

Answer:

According to the procedure to determine transport numbers in fused salts based on conductivity measurements (cf. Section 5.6.7.3 in the textbook), the first step is to determine the diffusion coefficient of the pair Na-Cl, i.e., $D_{Na\text{-}Cl}$. This can be calculated by applying the Nernst-Einstein equation to fused salts (cf. Eq. 5.76 in the textbook):

$$\Lambda' = \Lambda_{calc} - \frac{2\,zF^2}{RT} D_{Cs\text{-}Cl} \qquad\qquad (5.115)$$

or

$$D_{Cs\text{-}Cl} = \left(\Lambda_{calc} - \Lambda' \right) \frac{RT}{2\,zF^2} \qquad\qquad (5.116)$$

where Λ_{calc}, the calculated equivalent conductivity, is (cf. Eq. 5.75 in the textbook)

$$\Lambda_{calc} = \frac{zF^2}{RT}\left(D_{Cs^+} + D_{Cl^-}\right)$$

$$= \frac{(1)\left(96500\,C\,eq^{-1}\right)^2}{\left(8.314\,J\,mol^{-1}K^{-1}\right)(943\,K)}(3.5+3.8)\times10^{-5}\,cm^2\,s^{-1} \quad (5.117)$$

$$\times\frac{1J}{1CV}\times\frac{1V}{1A\Omega}\times\frac{1A}{1Cs^{-1}} = 86.7\,S\,cm^2\,eq^{-1}$$

Substituting Eq. (5.117) as well as the value of Λ' into Eq. (5.116)

$$D_{Cs-Cl} = \frac{(86.7-67.7)\,S\,cm^2\,eq^{-1}\left(8.314\,J\,mol^{-1}K^{-1}\right)(943\,K)}{2(1)\left(96500\,C\,eq^{-1}\right)^2} \quad (5.118)$$

$$\times\frac{1C^2\,\Omega\,s^{-1}}{1J} = 8.00\times10^{-6}\,cm^2\,s^{-1}$$

Once D_{Cs-Cl} is known, the next step is calculate the individual diffusion coefficients, $D_{Cs^+\,ind}$ and D_{Cl-ind}, of the independently jumping Cs^+ and Cl^- ions (cf. Eqs. 5.80 and 5.81 in the textbook):

$$D_{Cs^+\,ind} = D'_{Cs^+} - D_{Cs-Cl} = 3.5\times10^{-5}\,cm\,s^{-1} - 0.8\times10^{-5}\,cm\,s^{-1}$$
$$= 2.7\times10^{-5}\,cm\,s^{-1} \quad (5.119)$$

and

$$D_{Cl^-\,ind} = D'_{Cl^-} - D_{Cs-Cl} = 3.8\times10^{-5}\,cm\,s^{-1} - 0.8\times10^{-5}\,cm\,s^{-1}$$
$$= 3.0\times10^{-5}\,cm\,s^{-1} \quad (5.120)$$

Finally, the transport numbers are determined from the Einstein relation (cf. Eq. 4.172 in the textbook) and the relation between absolute and relative mobilities (cf. Eq. 5.84 in the textbook):

$$t_{Cs^+} = \frac{D_{Cs^+ ind}}{D_{Cs^+ ind} + D_{Cl^- ind}} = \frac{2.7 \times 10^{-5} \; cm \; s^{-1}}{(2.7 + 3.0) \times 10^{-5} \; cm \; s^{-1}} = 0.47 \quad (5.121)$$

and

$$t_{Cl^-} = 1 - t_{Cs^+} = 0.53 \qquad (5.122)$$

5.34 (a) Using the observed and calculated values of the equivalent conductivity at different temperatures given in Table 5.27 in the textbook for NaCl, find out the temperature dependence of the coordinated diffusion coefficient, i.e., the D_{Na-Cl} vs. T plot. **(b)** Determine the transport numbers of Na^+ and Cl^- in the molten salt. The self-diffusion coefficients of Na^+ and Cl^- in NaCl measured by the radiotracer method at 1113 K are $D_{Na+} = 9.60 \times 10^{-5}$ $cm^2 \; s^{-1}$ and $D_{Cl-} = 6.70 \times 10^{-5} \; cm^2 \; s^{-1}$. **(c)** The difference between calculated and observed equivalent conductivities is often phenomenologically attributed to the association (either permanently or transient) of cations and anions in the molten salt. What is the temperature dependence of this "association degree", and how would you explain the seeming contradiction with our knowledge about cation-anion interaction (cf. Problem 5.22)? (Cf. Problem 5.9 in the textbook) (Xu)

Answer:

(a) The diffusion coefficients measured by the radiotracer method include the contribution of the coordinated Na^+ and Cl^- pairs, i.e., D_{Na-Cl}. However, D_{Na-Cl} does not contribute to the current carrying flux. Therefore, in order to obtain the transport number, the independent diffusion coefficients of Na^+ and Cl^- are needed. From Table 5.27 in the textbook, the difference between the measured and the calculated equivalent conductivity can be used to derive the diffusion coefficients of the coordinated jump of Na^+ and Cl^-. Thus, using the values of equivalent conductivities at 1093 K, D_{Na-Cl} is (cf. Eq. 5.76 in the textbook)

$$D_{Na-Cl} = \frac{RT}{2zF^2}(\Lambda_{calc} - \Lambda') \qquad (5.123)$$

$$= \frac{\left(8.314 \, JK^{-1} mol^{-1}\right)(1093 \, K)}{2(1)\left(96500 \, C \, mol^{-1}\right)^2}\left(21 \, S \, cm^2 \, eq^{-1}\right) \times \frac{1 \, VC}{1 \, J} \times \frac{1 \, C\Omega s^{-1}}{1 \, V}$$

$$= 1.03 \times 10^{-5} \; cm^2 \, s^{-1}$$

In the same way for the other temperatures,

T (K)	1093	1143	1193	1293
Λ' (S cm^2 eq^{-1})	138	147	155	171
Λ_{calc} (S cm^2 eq^{-1})	159	177	198	240
Λ_{calc} - Λ'	21	30	43	69
$D_{Na\text{-}Cl}$ (cm^2 s^{-1})	1.03x10^{-5}	1.53x10^{-5}	2.29x10^{-5}	3.98x10^{-5}

Figure 5.13 shows the corresponding variation of $D_{Na\text{-}Cl}$ vs. T.

(b) The transport numbers for a 1:1 salt as a function of the corresponding independent diffusion coefficients can be calculated as (cf. Eq. 5.84 in the textbook),

$$t_i = \frac{D_i\left(ind\right)}{D_i\left(ind\right)+D_j\left(ind\right)} \tag{5.124}$$

To determine the independent diffusion coefficients of Na$^+$ and Cl$^-$, we make use of the equation (cf. Eqs. 5.80 and 5.81 in the textbook):

$$D_i\left(ind\right)= D_i' - D_{Na-Cl} \tag{5.125}$$

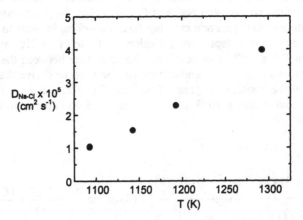

Figure 5.13. Temperature dependence of the coordinated diffusion coefficient of fused NaCl.

From Fig. 5.13, D_{Na-Cl} is found to have a strong temperature dependence. The value of D_{Na-Cl} at 1113 K can be obtained by a rough interpolation at this temperature, i.e., D_{Na-Cl} (1113K) $= 1.23 \times 10^{-5}$ cm^2 s^{-1}. Substituting the corresponding values into Eq. (5.125),

$$D_{Na^+}(ind) = D'_{Na^+} - D_{Na-Cl} = 9.60 \times 10^{-5} - 1.23 \times 10^{-5}$$
$$= 8.37 \times 10^{-5} \text{ cm}^2\text{s}^{-1} \tag{5.126}$$

and

$$D_{Cl^-}(ind) = D'_{Cl^-} - D_{Na-Cl} = 6.70 \times 10^{-5} - 1.23 \times 10^{-5}$$
$$= 5.47 \times 10^{-5} \text{ cm}^2\text{s}^{-1} \tag{5.127}$$

The transport numbers are, then, from Eq. (5.124),

$$t_{Na^+} = \frac{8.37 \times 10^{-5}}{8.37 \times 10^{-5} + 5.47 \times 10^{-5}} = 0.60 \tag{5.128}$$

and

$$t_{Cl^-} = 1 - 0.60 = 0.40 \tag{5.129}$$

(c) As shown in the table above, the difference of $\Lambda_{calc} - \Lambda'$ increases with temperature. This is the same as saying that the "association degree" rises with temperature. Furthermore, one would conclude that association between cation and anion is an endothermic process, which is favored as temperature rises. This contrasts our knowledge about ion interaction, because it is known that Coulombic attraction dominates the cation-anion pairing.

The reason for this discrepancy is that the "association" used here no longer involves the ion-pair formation in solvent-containing electrolytes, where both ions have "real" interactions and must stay for a relatively long time in the solvent sheath. In solvent-free electrolytes like molten salts, the "association" is much broader, including the paired jumps of both ions without interaction. When temperature goes up, the portion of "paired jump" increases as the total diffusion coefficient increases.

5.35 Consider the data in Figs. **5.50** and **5.51** in the textbook on diffusion coefficients as a function of temperature and pressure for the diffusion of ^{134}Cs ion in molten NaNO$_3$. Using these data find the heats of activation for the processes of *hole-formation* and *jump-into-a-hole*. Discuss whether they conform more to the *jump-into-a-hole* (Fürth) model, or the *shuffle-along* (Swallin) model of transportation. (GamboaAldeco)

Answer:

The heat of activation for the processes of *hole formation* ($\Delta H^{\neq}_H$) and *jumping-into-a-hole* ($\Delta H^{\neq}_J$) are given by the equations (cf. Eqs. 5.116 and 5.117 in the textbook),

$$-R\left[\frac{\partial \ln D}{\partial(1/T)}\right]_P = \Delta H^{\neq}_H + \Delta H^{\neq}_J \qquad (5.130)$$

and

$$-R\left[\frac{\partial \ln D}{\partial(1/T)}\right]_V = \Delta H^{\neq}_J \qquad (5.131)$$

From Fig. 5.51 in the textbook, the corresponding pressure and temperature values at a constant molar volume of, say, 45 cm^3 mol^{-1}, are :

	$V_m = 45$ cm^3 mol^{-1}			
T	350 ^{0}C (623 K)	370 ^{0}C (643 K)	390 ^{0}C (663 K)	420 ^{0}C (693 K)
P	0 atm	400 atm	730 atm	1200 atm

Using these pairs of data of temperature and pressure, the corresponding values of diffusion coefficient can be obtained from Fig. 5.50 in the textbook:

$1/T$ (K^{-1})	0.00161	0.00153	0.00151	0.00144
$[log_{10} D$ (cm^2 s^{-1})]$_V$	$\overline{5}$.365	$\overline{5}$.375	$\overline{5}$.385	$\overline{5}$.405
	= -4.635	= -4.625	= -4.615	= -4.595
$[\ln D$ (cm^2 s^{-1})]$_V$	-10.672	-10.649	-10.626	-10.580

The corresponding graph at constant volume is given in Fig. 5.14. The slope of the corresponding curve is

$$\left[\frac{\partial \ln D}{\partial(1/T)}\right]_V = -538.61\,\text{K} \tag{5.132}$$

Substituting this value into Eq. (5.131),

$$\Delta H_J^{\neq} = -\left(8.314\,\text{JK}^{-1}\text{mol}^{-1}\right)\left(-538.61\,\text{K}\right) = 4.478\,\text{kJ mol}^{-1} \tag{5.133}$$

Now, from the same Fig. 5.50 in the textbook, the diffusion coefficients at different temperatures and at a constant pressure of 800 atm are,

$1/T\,(K^{-1})$	0.00161	0.00153	0.00151	0.00144
$[log_{10}\,D\,(cm^2s^{-1})]_P$	$\bar{5}.265$	$\bar{5}.320$	$\bar{5}.380$	$\bar{5}.450$
	= -4.735	= -4.680	= -4.620	= -4.550
$[ln\,D\,(cm^2s^{-1})]_P$	-10.903	-10.776	-10.638	-10.477

Plotting $ln\,D$ vs. T at constant pressure gives the plot shown in Fig. 5.14. The slope of this curve is

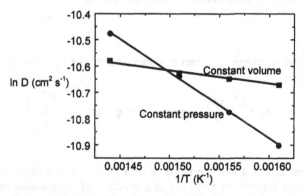

Figure 5.14. Plots of logarithm of the diffusion coefficient as a function of temperature at constant pressure and constant volume for the diffusion of ^{134}Cs ion in molten $NaNO_3$.

$$\left[\frac{\partial \ln D}{\partial(1/T)}\right]_P = -2520.88 \text{ K} \qquad (5.134)$$

Substituting this value as well as the value of $\Delta H_J^{\neq}$ from Eq. (5.133) into Eq. (5.130), and solving for $\Delta H_H^{\neq}$,

$$\Delta H_H^{\neq} = -\left(8.314 \text{JK}^{-1}\text{mol}^{-1}\right)\left(-2520.88 \text{K}\right) - 4478 \text{J mol}^{-1} \qquad (5.135)$$
$$= 16.480 \text{ kJ mol}^{-1}$$

If $\Delta H_J^{\neq} / \Delta H_H^{\neq} < 0.3$, then the *make-hole-then-jump-in* model makes sense. In this case, from Eqs. (5.133) and (5.135),

$$\frac{\Delta H_J^{\neq}}{\Delta H_H^{\neq}} = \frac{4.478 \text{kJ mol}^{-1}}{16.480 \text{kJ mol}^{-1}} = 0.271 < 0.3 \qquad (5.136)$$

which means that the *make-hole-then-jump-in* model better describes the process of transportation of ^{134}Cs ion in molten $NaNO_3$.

5.36 Figures 5.29 and 5.68 in the textbook show two plots of viscosity as a function of temperature for a molten salt and liquid SiO_2. Determine the energy of activation for these two systems. What are the relative diffusion coefficients at their melting points considering the D_0's to be approximately equal? Comment on your results. (GamboaAldeco)

Answer:

The activation energy, $E_\eta^{\neq}$, can be obtained from the equation,

$$\eta = \eta_0 e^{E_\eta^{\neq}/RT} \qquad \text{or} \qquad \ln\eta = \ln\eta_0 + \frac{E_\eta^{\neq}}{R}\frac{1}{T} \qquad (5.137)$$

This equation indicates that the slope of a *ln* η vs. *1/T* plot is equal to $E_\eta^{\neq}/R$. Therefore, from the plot of viscosity as a function of temperature in the range 1000 to 1250 K for the molten salt in Fig. 5.29 in the textbook, $m = E_\eta^{\neq}/R = 2500$ K or

$$E_\eta^{\neq} = (2500\,K)\left(8.314\,JK^{-1}mol^{-1}\right) = 20.79kJ\,mol^{-1} \qquad (5.138)$$

In the same form, from the slope of the plot for SiO_2 in Fig. 5.68 in the textbook in the range 2200-2300 K, , $m = E_\eta^{\neq}/R = 29500$ K   or,

$$E_\eta^{\neq} = (29500\,K)\left(8.314\,JK^{-1}mol^{-1}\right) = 245.26kJ\,mol^{-1} \qquad (5.139)$$

Considering $E_\eta^{\neq} \approx E_D^{\neq}$ (cf. Eq. 5.60 in the textbook), then, the self-diffusion coefficient can be approximated as,

$$D = D_0\,e^{-E_\eta^{\neq}/RT} \qquad (5.140)$$

If $(D_0)_{m.s.} \approx (D_0)_{SiO_2}$, then,

$$\frac{D_{m.s.}}{D_{SiO_2}} \approx \frac{\exp\left[-\left(E_\eta^{\neq}\right)_{m.s.}/RT_{m.s.}\right]}{\exp\left[-\left(E_\eta^{\neq}\right)_{SiO_2}/RT_{SiO_2}\right]} = \exp\frac{-\left(E_\eta^{\neq}\right)_{m.s.}}{RT_{m.s.}} + \frac{\left(E_\eta^{\neq}\right)_{SiO_2}}{RT_{SiO_2}}$$

$$(5.141)$$

Evaluating Eq. (5.141)

$$\frac{D_{m.s.}}{D_{SiO_2}} \approx \exp\left[-\frac{20790J}{\left(8.314\,JK^{-1}mol^{-1}\right)(1074K)}\right.$$

$$(5.142)$$

$$\left.+\frac{245260J}{\left(8.314\,JK^{-1}mol^{-1}\right)(2073)}\right] = 1.5 \times 10^5$$

This equation indicates that the diffusion coefficient of the molten salt is about 10^5 times greater than that of the liquid SiO_2.

Comment: The diffusion coefficient is directly related to the mobility of the species (cf. Eq. 4.172 in the textbook). The mobility and, therefore, the specific conductivity of the molten salt are expected to be 10^5 times greater than that of the liquid SiO_2 (cf. Table 5.44 in the textbook). The big difference in conductivities is because molten salts are constituted by ionic lattices that upon melting form highly conductive *ionic* liquids, in contrast to SiO_2 that fusses into an *associated* liquid.

5.37 Assuming a 3-coordination exclusively for B as well as a planar structure in borate glasses (a) determine the formula of the glass, and (b) calculate the moles of base (e.g.Na₂O) needed to form a chain structure. (Xu)

Answer:

The three coordinated boron has all its bonds on the same plane, therefore the borate glass has a sheet-like structure as shown in Fig. 5.15. The Lewis center, B^{+3} in this case, can be attacked by oxygen anions and the network gets ruptured as shown in Fig. 5.16. It can be seen that one mole of Na_2O breaks one mole of B-O bonds. To transform the 2D sheets into 1D chain, one out of the three B-O bonds must be broken for every boron center, i.e., the ring has to be broken in one point. Since 1.0 *mole* of pure borate, B_2O_3, has

$$3N_B = 3(1mol)(6.022 \times 10^{23}) = 1.8 \times 10^{24} \text{ B-O bonds} \qquad (5.143)$$

Therefore for every mole of borate to form a 1D chain, 1 mole of B-O bonds need to be broken, and the amount of Na_2O needed is **1 mole**. So, the component of 1D borate glass should be:

$$B_2O_3 + Na_2O \rightarrow Na_2B_2O_4 \qquad (5.144)$$

and its formula, **NaBO₂.**

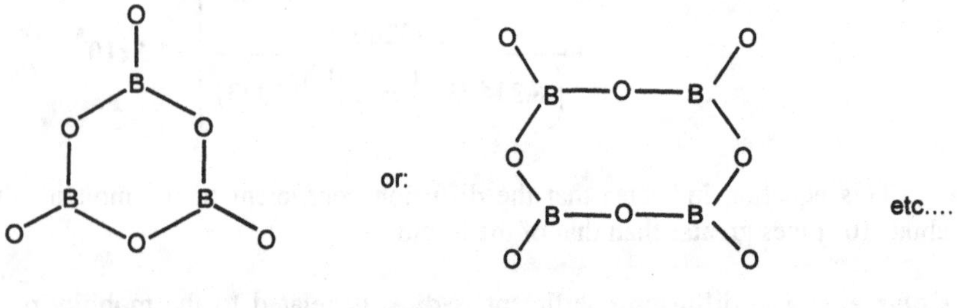

Figure 5.15. Planar structure of Na₂O considering only a 3-coordination.

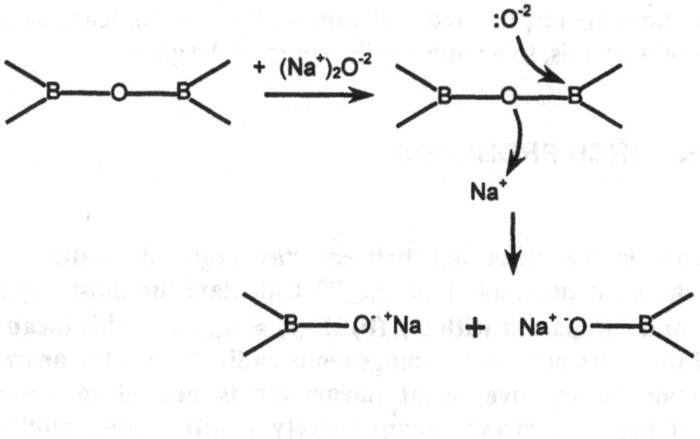

Figure. 5.16. Attack of Na_2O to borate glass.

5.38 According to Adam-Gibbs theory, the constant B in the non-Arrhenius equation for viscosity (cf. Problem 5.31) is given by,

$$B = \frac{\Delta\mu S_c^*}{kC}$$ (5.145)

where $\Delta\mu$, S_c^*, k and C are the activation energy for the transport process, the configurational entropy of the subsystem, the Boltzmann constant, and a proportionality constant in a heat capacity-temperature relation, respectively. As found in Problem 5.31, the value of B determines the fragility of the liquid, which can be classified into three categories: *strong* (large value of B), *intermediate* (medium value of B), and *fragile* (small value of B). Pure silicate belongs to the *strong* class, with a B value of ca. 100. As Na_2O is added, the fragility increases and the resultant glass passes via *intermediate* ($B < 50$), to *fragile* ($B < 10$). Interpret this transformation on a structural level. (Xu)

Answer:

As Na_2O is added to the pure silicate, the Si-O bonds are broken up, and large chunks of silicate are disintegrated into relatively small pieces with either chain or ring structure. When the structural unit becomes smaller, so does the sub-system whose cooperation is required for the transportation process to happen in the Adam-Gibbs theory. Hence, S_c^*, the configurational entropy, becomes smaller. On the other hand, as the network is being broken up, viscosity drops, either holes needs much smaller activation energy to form, or silicate pieces of certain sizes need smaller activation energy to jump into the formed hole. Both effects, i.e., reduced values of S_c^* or $\Delta\mu$, leads to smaller B values, or in other words, to a more fragile nature of the glass.

MICRO-RESEARCH PROBLEMS

5.39 (a) What is the difference between "average-hole radius, $\langle r \rangle$," and "radius of the most populous hole, r_{max}"? Calculate the most-popular-hole radius r_{max}, and compare it with $\langle r \rangle$. (b) If $\langle r \rangle \approx r_{max}$, does this mean that the majority of the holes possess a homogeneous radius? (c) If the answer in the above question is negative, what parameter is needed to describe the dispersity of the hole sizes? Quantitatively confirm the validity of the approximation that "all holes are of the same size in molten salts". Using the data in Table 5.15 in the textbook, find the above dispersity for KCl molten salt at 900 ^{0}C. [Hint: Numerical integration may be needed to solve question (c)]. (Cf. Micro Research 5.2 in the textbook) (Xu)

Answer:

(a) The most popular radius, r_{max}, refers to the radius of the holes whose number is larger than the number of holes of any other radius. This number does not necessarily (and often definitely not!) stand for majority compared with the total number of holes present. In other words, the probability takes a maximum at this radius (see Fig. 5.13). The value of r_{max} differs from that of $\langle r \rangle$ unless the distribution of r is roughly symmetrical against r_{max} (Fig. 5.14) Unlike $\langle r \rangle$, r_{max} is not an average value. Since a maximum in the probability corresponds to r_{max}, this parameter can be derived from the general procedure for obtaining an

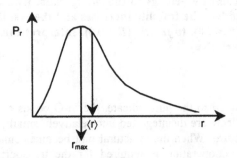

Figure 5.13. Plot of the probability against the hole radius, showing the difference between "average hole radius" and the radius of the "most populous hole".

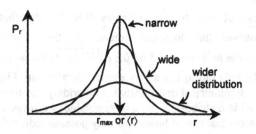

Figure 5.14. Plot of the hole probability against the hole radius. When the distribution is symmetrical, then $\langle r \rangle = r_{max}$.

extremum as described next. The probability of having a radius between r and $r + dr$ is known to be (cf. Eq. 5.39 in the textbook):

$$P_r \, dr = \frac{16}{15\sqrt{\pi}} a^{7/2} r^6 e^{-ar^2} \, dr \tag{5.146}$$

Differentiating P_r with respect to r, one gets,

$$\frac{d}{dr} P_r = \frac{16}{15\sqrt{\pi}} a^{7/2} \left(6r^5 e^{-ar^2} - 2ar^7 e^{-ar^2} \right)$$
$$= \frac{16}{15\sqrt{\pi}} a^{7/2} \left(6r^5 - 2ar^7 \right) e^{-ar^2} \tag{5.147}$$

Equating Eq. (5.147) to zero and solving it for r,

$$r_{max}^2 = \frac{3}{a} \tag{5.148}$$

Inserting now the value of the parameter a (Eq. 5.35 in the textbook), i.e., $a = 4\pi\gamma/kT$,

$$r_{max} = \sqrt{\frac{3kT}{4\pi\gamma}} = 0.49 \sqrt{\frac{kT}{\gamma}} \tag{5.149}$$

Eq. (5.149) represents the most populous-hole radius, which is very close to the average radius, $\langle r \rangle$ of $0.51\sqrt{kT/\gamma}$ derived in Eq. (5.44) in the textbook.

(b) The closeness of r_{max} to $\langle r \rangle$ indicates that the distribution is almost symmetrical. However, this does not indicate that the *majority* of the holes possess a radius in the neighborhood of $0.51\sqrt{kT/\gamma}$. This would be equivalent to thinking that the hole radius has a very narrow distribution. Figure 5.14 shows the comparison of symmetrical distributions with widely and narrowly dispersed hole-size. In the wide distribution, the average value is always somewhere near r_{max}, although a large quantity of holes still may possess radii either much larger or smaller than $\langle r \rangle$. On the other hand, in the narrow distribution all the holes have approximately the same size. A wider dispersed hole-size distribution may have the same average properties than the narrow dispersed distribution, but its holes possessing radii close to $\langle r \rangle$ are far from being majority. In this case, the approximation that "all holes are of the same size" will not hold true. Therefore, the closeness of r_{max} to $\langle r \rangle$ is not enough to guarantee a narrowly dispersed hole-size around $\langle r \rangle$.

(c) There may be more than one way to solve this problem, and the following is only a possible path. It is already known, by definition, that the total probability for holes of all possible sizes is unity, which is the whole area under the curve P_r vs. r, i.e.,

$$\int_0^\infty P_r\, dr = 1.0 \qquad\qquad (5.150)$$

To actually reflect the dispersity of the hole sizes, one needs to know the probability of existence of holes of sizes "around" the most populous radius r_{max}. In other words, one needs to find out the radius variation, Δr, that makes the probability for holes of radii between $r_{max} - \Delta r$ and $r_{max} + \Delta r$ be, say, 50% of the total probability for holes of all possible sizes. In this way, the larger Δr is, the more dispersed is the hole radius distribution. Now that the hole dispersity is defined as Δr, the problem becomes to solve Δr from the following equation:

$$\frac{\int_{r_{max}-\Delta r}^{r_{max}+\Delta r} P_r\, dr}{\int_0^\infty P_r\, dr} = \frac{\int_{r_{max}-\Delta r}^{r_{max}+\Delta r} P_r\, dr}{1.0} = 0.5 \qquad\qquad (5.151)$$

The integration $\int_{r_{max}-\Delta r}^{r_{max}+\Delta r} P_r\, dr$ can only be evaluated through numerical methods. The value of P_r can be obtained from Eq. (5.39) in the textbook, i.e.,

$$P_r = \frac{16}{15\sqrt{\pi}} a^{7/2} r^6 e^{-ar^2} \tag{5.152}$$

KCl has a surface tension of 89.5 dyne cm^{-1} (cf. Table 5.15 in the textbook.) From Eq. (5.149), r_{max} at 900 ^{0}C is

$$r_{max} = 0.49\sqrt{\frac{kT}{\gamma}} = 0.49\sqrt{\frac{\left(1.38 \times 10^{-16}\, \text{ergK}^{-1}\right)(1173\,\text{K})}{89.5\,\text{dyne cm}^{-1}}} \tag{5.153}$$

$$= 2.08 \times 10^{-10}\, \text{m}$$

With this data, Eq. (5.151) can be written as,

$$\int_{r_{max}-\Delta r}^{r_{max}+\Delta r} P_r\, dr = 1.68 \times 10^{69} \int_{r_{max}-\Delta r}^{r_{max}+\Delta r} r^6 e^{-ar^2}\, dr = 0.5 \tag{5.154}$$

By trial-and-error, Δr is found to be 0.4×10^{-10} m. This result means that the radius of 50% of the holes, is in between 1.68×10^{-10} m and 2.48×10^{-10} m. This corresponds to a very narrowly dispersed distribution (Fig. 5.15) since Δr is smaller than half of the size of the studied ion. A relation between various Δr and $\int_{r_{max}-\Delta r}^{r_{max}+\Delta r} P_r\, dr$ is plotted in Fig. 5.16. This figure shows that as Δr increases

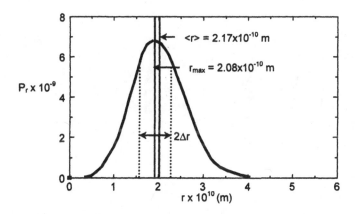

Figure 5.15. Plot of the hole probability against the hole radius for fused KCl at 900 ^{0}C.

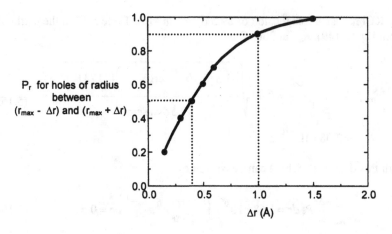

Figure 5.16. Plot of $\int_{r_{max}-\Delta r}^{r_{max}+\Delta r} P_r\, dr$ against Δr for fused KCl at 900 °C.

from 0.4×10^{-10} m to 1.0×10^{-10} m, the integration rapidly increases from 0.50 to 0.91. Therefore, the approximation that all the holes present have the same radius is effective and close to truth.

INDEX

9 780306 466687